What people are saying about

Reflections on a Surprising Universe

In clear prose, Richard Dieter lays out a vision of the universe filled with curiosity and wonder. Dieter's layman vision of the universe brings the majesty of the cosmos down to the level of the everyday in census of the weirdness that surrounds us on all sides. From the smallest particles to the vastest super-structures, this book is a great read for anyone looking up in the sky and wondering what's out there.
John Wenz, former associate editor, *Astronomy* magazine

Not many authors have the skill and breadth to masterfully explain the mysteries of science, and do so in literate language, but Richard Dieter easily ranks as one of them.
Colman McCarthy, former *Washington Post* columnist

T0163437

Reflections on a Surprising Universe

Extraordinary Discoveries
Through Ordinary Eyes

Reflections on a Surprising Universe

Extraordinary Discoveries
Through Ordinary Eyes

Richard Dieter

IFF
BOOKS

Winchester, UK
Washington, USA

JOHN HUNT PUBLISHING

First published by iff Books, 2019
iff Books is an imprint of John Hunt Publishing Ltd., No. 3 East Street, Alresford,
Hampshire SO24 9EE, UK
office@jhpbooks.com
www.johnhuntpublishing.com
www.iff-books.com

For distributor details and how to order please visit the 'Ordering' section on our website.

Text copyright: Richard Dieter 2018

ISBN: 978 1 78904 202 3
978 1 78904 203 0 (ebook)
Library of Congress Control Number: 2018961974

A CIP catalogue record for this book is available from the British Library.

Design: Stuart Davies

UK: Printed and bound by CPI Group (UK) Ltd, Croydon, CR0 4YY
US: Printed and bound by Thomson-Shore, 7300 West Joy Road, Dexter, MI 48130

We operate a distinctive and ethical publishing philosophy in
all areas of our business, from our global network of authors to
production and worldwide distribution.

Contents

Preface

If you picked up this book, you're probably wondering what you might find inside. To start with, this book is about how our universe appears from the perspective of science. The "surprise" in the title comes from the revelation that our world is vastly different from the way it appears at first glance. The far reaches of space are vaster, more complex, and more explosive than one might imagine from just gazing up at the night sky. Looking within, we find that the fundamental building blocks of the universe—at least the ones we have discovered—defy all explanation in ordinary terms. They erase our common notions of solidity, location, and determinability. Yet we have come to understand this amazing place at a level that has enabled us to reach both far beyond our humble planet and deeply into the world of invisible particles and forces.

Hopefully, however, this book will be more than a collection of discoveries—amazing though those discoveries may be. Reading this book is meant to be an enjoyable and broadening experience. I have tried to translate the rigors of scientific breakthroughs into a readable narrative. I have endeavored to connect the strings of the vast cosmos to the elementary building blocks of our existence, while all the time relating them to the human scale in which we live.

For those who have not taken this journey before, as well as for those who relish taking it again from a fresh perspective, I hope this book will convey a sense of wonder, like peering through a window to a beautiful new vista that has been hidden by clouds. Accepting that our home planet is but a speck of dust in a universe that seems unconcerned about our existence, while still feeling pride for the knowledge we have and will acquire about how this all works, will hopefully also affect how we care for our planet and for each other.

There are many places a reader can go for more in-depth accounts of the science discussed here. The Bibliography and the Notes offer some of those avenues. I suppose it would be possible to reference every sentence with another source and to give more credit where it is due. Instead, I have used the Notes to indicate where a broad topic is more thoroughly discussed in another book or article, especially when that source has helped my own understanding. I wish to thank Walter Liggett, former Mathematical Statistician, National Institute of Standards and Technology, and John Wenz, former associate editor, *Astronomy* magazine, for reading through the manuscript and making helpful suggestions. When errors arise, however, they are my responsibility.

– *Richard Dieter*

Part I

Looking for Our Place in the Stars

The laws of nature form a system that is extremely fined-tuned, and very little in physical law can be altered without destroying the possibility of the development of life as we know it.

– Stephen Hawking[1]

Introduction

This is a book about where we fit in the vastness of the universe. Scientific knowledge is growing at an ever-increasing pace, yet much of it seems to have little connection to our daily lives. These pages examine some of science's amazing discoveries, while also pausing to ask, "What does this mean for us?" Science is not only out there but also in us. As the only known life form in the universe contemplating the mysteries that surround us, a look beyond our daily routine could be enriching. There's always a human dimension to things we learn about our world, even if the insight is how insignificant the human dimension appears to be.

Experimental science is only about five hundred years old, a small fraction of our time as a species on Earth. Ancient civilizations had theories about the structure of the universe and where we came from, but there was little technology to test those theories. The expansion of scientific knowledge has gone hand-in-hand with the advancement of the tools to explore it. Our telescopes can now see almost back to the beginning of time, our microscopes are piercing through to the very atoms of our existence. We have machines that come close to duplicating the extreme conditions when our universe was smaller than the size of a tiny thimble. Unless we are somehow blown back to a more primitive state in which survival again becomes the only goal, science will continue to make more astounding discoveries, many of which will ultimately change our lives and our future. Surprisingly, it is possible for us to make sense of much of what goes on in this vast cosmos, even to the point of approximating some of its laws.

People have always been intrigued by great discoveries in science, even if the implications of those findings were unsettling to some. Early scientists like Giordano Bruno and

Galileo Galilei were punished for promulgating theories that contradicted ecclesiastical teachings about creation and the centrality of humans in the world. But ultimately, the truths of their discoveries were embraced because they could be verified. The high regard for great thinkers from Isaac Newton to Albert Einstein is a measure of the increasing appreciation for the role of science in society, even as nagging misunderstandings can still lead to resistance.

The recent fascination with the solar eclipse in 2017, which was visible from coast to coast in the United States, shows that scientific wonder is alive and well. However, some scientific discoveries go so far beyond the human scale that they seem to be incomprehensible and irrelevant to the average person. Compared to the moon blocking the sun, it is difficult for most people to see much significance in the collision of two black holes trillions of miles away, whose vibrations we're just recording as gravitational waves billions of years after the event occurred.

True science has always been a mixture of theory and experiment. Sometimes the theory comes first, as with Einstein's conjecture that a small amount of matter could be converted into an enormous amount of energy, or that time actually slows down when you travel fast, compared to a stationary observer. These theories were only confirmed long after they were proposed, thereby establishing Einstein's status as a genius.

At other times, observation and experiment force a modification of existing theory, as with Edwin Hubble's measurements that the universe appeared to be rapidly expanding, challenging the notion that it was in a steady state. Most recently, we discovered that almost all the energy and matter in the universe are unaccounted for by present models and are unlike anything we know. Theories to adequately explain this "dark energy" and "dark matter" have yet to be devised.

Given our present knowledge, there's little that we can say with certainty about our relationship to the universe. It is just too big and mysterious a place. As things stand now, humans could be just an aberrant twig on a small branch of mammals, leading out from an immense trunk of evolutionary life that we don't fully comprehend. And there might be other trees of life on other planets, as well. We could be one of nature's short-lived experiments that is quickly extinguished in favor of more long-lasting life forms. Even life itself may be just a sideshow among many other complex forms of existence, but is otherwise of little consequence compared to the main event taking place in the continuous birth of stars and galaxies throughout the universe.

It would be sad, but instructive, if we found that life frequently occurs in the universe but that it is usually snuffed out when its complexity allows for a self-conscious species. We might find that wherever temperatures are moderate and chemicals abundant, sooner or later a certain combination of molecules will reproduce itself. The ones that survive will be the ones that absorb other chemicals for energy, and reproduce again and again, evolving into more intricate forms. But when complexity leads to dominance by a species capable of affecting the entire planet, life itself might come to a halt.

Evolution could be taking place on many worlds, moving from simple organisms to the more advanced, until one creature has the ability to wonder how their world was made. Sooner or later, such creatures are bound to discover the key relationship between matter and energy, believed to be true throughout the universe: $E=mc^2$. The conversion of matter into energy that sustains civilizations through heat from the sun can also be used to cause catastrophic destruction. Once that relationship is discovered, it could be that the demise of intelligent life takes place very quickly. If any life remains, it is primitive, until perhaps the cycle starts over again. Discovering such a pattern among other forms of life would be a grave warning that we better take steps to avoid a similar end.

On the other hand, humans might be the key link toward a new species with incredible intelligence and influence in the future. In humans, life made a giant leap from the millions of species of plants and animals that have come and gone. We appear to be the first on Earth to have a vibrant self-consciousness, complex languages and forms of communication that record history and enable new discoveries. Each species is unique, but humans seem to occupy a whole other level of creativity, making arguably a greater leap than that from inanimate to animate. Humans have done amazing things in a relatively short time, and seem to be capable of infinitely more, if we manage to remain in existence.

Beyond the limits of science, it is impossible to rule out that humans are much more than an extraordinary example of what happens with significant brain expansion, and in fact rather represent the full flowering of a creation, born and intended to be in a unique relationship with the mover behind the universe, possessed of a spiritual side that transcends the material. This book, however, will stay tethered to what we can know and measure, leaving more theological questions to others. Science— even on the most basic level addressed in this book—will be hard enough to explain and understand.

Science does not reveal its mysteries easily. We're happy when technology offers us a brilliant new application that makes our lives more exciting and easier. Cures for diseases, flights into outer space, computers and the Internet are all triumphs of science and engineering. But to truly understand the world at its most elementary level, or to probe the universe as a whole and where it is headed, you would need to choose a scientific specialty and pursue years of study. Understanding science today requires us to imagine things that are beyond our power to visualize or even express in ordinary terms, such as a world with more than three dimensions, or one in which things have no definite location, or even where whole other universes can emerge in a fraction of a second.

In short, we live in a world that is not always of human scale and within our language or experience. Yet, we keep trying to understand it. We admire the scientists who spend their lives probing the remotest corners of knowledge, even if we know we can never join them.

In one way, it's surprising that any of our theories hold up and our proposed equations are found reliable. Why should the universe conform fairly closely to the laws dreamed up by one species on the obscure planet Earth? The laws governing the cosmos existed billions of years before our planet ever came to be. It's as if a secret vault exists containing the physical principles that govern the universe. Occasionally, we stumble upon one of these laws through our thoughts and experiments, but they were always true with us or without us.

Some of today's science includes sweeping theories that may only be tested indirectly, if at all. Our universe may have emerged from a tiny bundle of energy, but we may never be able to go back to the beginning of time to observe the big bang. And how do we explore what happened before then if all our scientific laws don't apply? All matter and energy may be made up of tiny one-dimensional strings, much smaller than atoms and probably unobservable, whatever our technology. Matter apparently can pop out of the vacuum of nothingness, and other universes—permanently outside of our reach—may be forming all the time—or maybe not.

Why should we bother to explore science at all? I believe that is where the beckoning of the stars comes in. What draws us to take on new challenges could be our relationship to the stars. We are literally made of stardust, the Earth is only our present home. Despite our human limitations, we have always looked to the stars. We are all scientists with a bit of wonder. We want to know more.

One thing seems clear. We will have to adapt our way of thinking and talking about the world to fit the reality that presents itself. Quantum computers will almost surely be a tool of the future, even though almost no one understands what a quantum is. Gravitational waves may be the telescope of tomorrow, even as we still tend to rely on Newton's notion that the Earth attracts apples. Future humans will probably have no trouble thinking of space in more than three dimensions and will feel comfortable calculating how much younger they will be after traveling near light speed compared to those who stayed on the ground.

Out there may be our future, or there may be civilizations that understand the universe in a way that can be explained in one great theory. If we can hold ourselves together as a species, our journey to the stars—and the other planets they possess—seems inevitable. It's worth trying to understand with how far along the journey we have already progressed.

Chapter 1

Exploring the Universe on Its Grandest Scale

At every step, the expansion of our universe was at the same time an expansion of the human horizons of knowledge.
– Mario Livio[1]

Humans have always gazed up into the night sky and wondered: what else is out there? The daytime is our comfort zone with just one star—our sun—reliably warming us and sustaining life. But at night, the stars are aloof, twinkling against a black sky, withholding their mysteries. Could there be other places in the universe, maybe other green and blue planets where life is also thriving, perhaps looking back down at us?

What do we know about this immense universe that reveals itself only grudgingly? The stars barely change their orientation from night to night and stay fixed with respect to each other. The big dipper roams the sky as the Earth turns, but it is always a dipper. The stars reveal almost nothing of themselves. Even with a telescope, they may only appear as bright dots.

If you watch the sky carefully, you can't help but notice that some points of light have an independent streak. They change their position more dramatically than the other lights, moving right through the constellations instead of staying with one. These are the planets of our solar system.

The Gifts of Jupiter

In a backyard telescope, it's possible to explore a few of these planets more closely. My favorite one is Jupiter. It's very bright and easy to see, although it just looks like a bright star to the naked eye. But with a telescope, a whole new world appears.

Jupiter has color, bands of yellow and brown. The bands encircle the planet and appear tilted at an angle relative to our perspective. It's not hard to picture that the planet is revolving around its axis and coursing through space as the night passes. Something else pops into view with a telescope centered on Jupiter. There are four small lights close to the planet. Depending on when you look, you might only see two or three, but eventually all four will appear. Sometimes they seem to be in a straight line. Clearly, they are associated with the planet because they follow it through the sky. They are four of the many (79 in all) moons of Jupiter, constantly orbiting the planet just as our moon orbits the Earth. Sometimes one passes in front of the planet (relative to our line of sight), and you can see its shadow as a dark dot on the planet's surface. At other times the moons pass behind the planet, decreasing the number of moons you can see.

Jupiter is a stunning example that our universe may be full of other worlds, different from the Earth, but not like the fiery and distant stars. Jupiter has contributed to important scientific discoveries that have changed our understanding of the universe. In the 17th century, Galileo Galilei was one of the first persons to explore the sky with a telescope. He was able to see the same moons of Jupiter that an amateur astronomer can easily see today. He realized that if those four moons were orbiting Jupiter, then they are not simply part of a dome of lights that encircles the Earth and moves across the night sky, as was originally thought. The moons of Jupiter were one piece of evidence that the Earth may not be the center of the universe. Gradually, it became clear that the Earth and all of our neighboring planets orbit the sun, ending the assumptions about the Earth having an exceptional position in the sky.

Jupiter and its moons led to another critical breakthrough—measuring the speed of light. Light is essential to everything humans do. Although it is impossible for our senses to detect the movement of light, there were reasons to believe that it did not appear instantaneously, but moved through space like sound travels through the air, though much more quickly.

The reason we are able to see Jupiter and its moons is that the light of the sun illuminates them and that light is reflected back towards Earth. (That is not true of the stars, which are their own source of light.) The distance that light has to travel in reaching the Earth from Jupiter varies depending on whether it is relatively nearer or further from us. Think of two runners on an oval track running at different speeds. At first, the faster runner will pull away from the slower one, but eventually she will lap the slower runner and the two will be close again. Jupiter takes 12 Earth-years to complete a trip around the sun, while the Earth, of course, does it in one year.

The moons of Jupiter orbit it in a regular fashion, so once a moon, say Io, disappears behind Jupiter in its orbit, its reappearance will occur at a predictable time. But if Jupiter and its moons are at a period when they are farther away from Earth, the light from the emerging moon will take longer to reach the Earth than if this was all happening at a time when Jupiter was

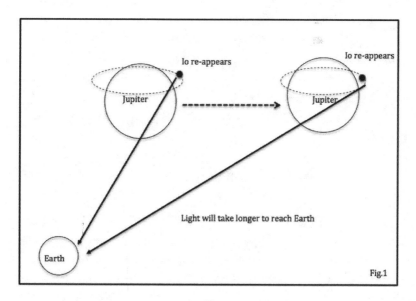

closer to Earth. If we know the difference in time it takes for Io to reappear and the difference in distance between the far and close observing times of Jupiter, we can approximately calculate the speed of light. Speed is simply distance divided by time. Using this idea, the first credible measurement of the speed of light was made by Ole Rømer in 1676. His data led to a calculation of 131,000 miles per second, at least in the ballpark of the actual figure of 186,300 miles per second.[2]

(I recently read of an even simpler experiment involving only Velveeta cheese and a microwave oven that could also be used to calculate the speed of light. Unfortunately, no one in the 1600s had a microwave oven, or Velveeta cheese. They also didn't know that light traveled in waves, which is critical to this very rough calculation.)[3]

The speed of light turns out to be one of the most important numbers describing the world we live in. It's a universal speed limit and is surprisingly independent of whether light comes from an object moving towards you or away. More on that later. (By the way, Saturn with its rings and moons is also an amazing sight through a backyard telescope.)

Probes and a few rovers have explored all of the planets in our solar system, and these places seem unlikely to sustain life like ours. In the near future, we will probably know if our neighbor Mars has supported microscopic life, perhaps frozen in its underground ice. More promisingly, there may be existing life in the oceans of one of the moons of Jupiter or Saturn. There is much to explore.

The Life of Stars

The more distant stars gradually revealed their secrets to astronomers as larger telescopes were developed. How far away are the stars? Do they orbit around anything else? What are they made of? How many are there? If they are very far away, but still glow with an evident brightness, perhaps they are hot furnaces like our sun. We cannot feel their warmth, but some source of energy is sending their light across the universe in a constant glow.

We now believe that the distant stars and our sun are indeed similar. There are thousands of stars visible to the naked eye, with a concentrated band of them glowing across the sky in what is called the Milky Way. And if stars are like our sun, perhaps they, too, have planets orbiting them, thereby offering a vast number of the potential places where life could exist in the universe. Although planets around other suns have not been seen directly, there is strong evidence that solar systems like ours are quite common.

The closest other sun (star) to us is Proxima Centauri. Its light takes 4.24 years to reach us, traveling a distance of about 25 trillion miles. By comparison, our own sun is "only" about 93 million miles away from the Earth. Distances in space quickly become too cumbersome to express in miles, so astronomers came up with a more manageable metric, and then gave it a confusing name: light-years. It sounds like an amount of time, but actually is the *distance* that light travels in an Earth year. It is equivalent to approximately 5.9 trillion miles, with light traveling at 670 million miles per hour. (Scientifically speaking, these distances should be expressed in meters.) Proxima Centauri, therefore, is 4.24 light-years away from us (25 trillion miles away ÷ 5.9 trillion miles per light-year). If a giant flare erupted there, we wouldn't see it for over 4 years. A similar flare on our own sun wouldn't be visible to us for about 9 minutes.

The rest of the stars are further away, some much, much further. On a clear night, away from the city, you might be able to observe some fuzzy white puffs in the sky, less defined than the point-like stars. Originally, these were called nebula because they resembled a gaseous collection of light. More powerful telescopes detected that some of those fuzzy clouds were actually conglomerations of stars.[4] Suddenly the universe became immensely larger.

Large collections of stars bound together by gravity are called galaxies. All the distinct stars our unaided eyes can see by gazing upward are part of our Milky Way Galaxy, which would also appear as a fuzzy blob if seen from very far away. The discovery in the early 20th century that other galaxies existed opened up vast possibilities for the existence of other life in the universe. Each galaxy contains billions of stars, and each star could be surrounded by a set of planets. Just as the Earth is not the center of our solar system, our solar system is not the center of our galaxy, and our galaxy is just one of hundreds of billions of galaxies in the universe.

Typical galaxies contain at least 100 billion stars, with larger ones encompassing a trillion stars. There are approximately 200 billion such galaxies in our known universe.[5] Hence, our sun is just one of about 20,000,000,000,000,000,000,000 other stars. At times it feels like we barely matter. It's hard to imagine that the vast universe cares about whether we are happy or sad. This is certainly the sentiment of the universe's bureaucracy in *The Hitchhiker's Guide to the Galaxy* by Douglas Adams,[6] where the Earth (and all its inhabitants) is scheduled for demolition because it stands in the way of a proposed interstellar highway in space. We're just a boulder along a future road that needs to be cleared away.

On the other hand, the universe constantly connects with us in subtle ways. When two black holes over a billion light-years away collide, they send out a signal, a ripple in the fabric of spacetime, that we humans can detect. (See Chapter 2.) We may not be the center of this vast world, but that is partly because the universe has no center. It is all interconnected.

Big, and Getting Much Bigger

If the number of stars weren't daunting enough, it appears that our whole visible universe is expanding. Careful measurements of the most distant galaxies reveal that they are retreating from us, and the further ones are retreating faster than the closer ones. Since the stars and the galaxies are so far away and there is no reference point to measure how far they have traveled, it's difficult to determine their speed. Light travels in waves, and if the source of the light is receding from us, the waves reach us in a more stretched-out form. The same thing happens to sound waves, which is why an ambulance siren moving away from you has a different pitch than when it is approaching you. Astronomers have been able to measure the relative stretching of light waves and have thus arrived at the speed at which stars and galaxies are receding.[7]

Are the stars and galaxies spreading out into the infinite emptiness of space? Not exactly. It is space that is actually expanding, though that is a hard notion to get one's head around. Scientists don't really know what space is. They are adamant that it's not empty nothingness.[8] It might be lumpy at the finest-grained level, or it might be continuous. It contains fields of energy and virtual particles popping into and out of existence. We will explore the really small aspects of the universe in the second section of this book. Suffice for now to say that the container of all our stuff is growing from the inside out. It's not just stars at the furthest limits that are moving away from us, it's every bit of space from our solar system to the space between the farthest galaxies. It is growing like a rising muffin in the oven, expanding from the inside.

Why haven't we noticed this before? One reason is that there are counter forces that keep the world closest to us looking like itself, even as distant galaxies are receding from ours at vast speeds. On a molecular level, forces within atoms keep us together so that our waistlines (and heads) don't grow with the expanding universe. In every atom, protons attract electrons, and even forces within the nucleus of an atom bind its particles together.

The Earth is partly held together by gravity, as is our solar system and our Milky Way Galaxy. Even clusters of galaxies are bound by the mutual attraction of their members. That muffin of space in the oven is really a giant blueberry muffin, just starting to bake. It is rising all around us, but the blueberries remain intact, even as they drift further apart from one another.

More recently, another surprising discovery was made. The universe is not just expanding, the rate of expansion is increasing.[9] It's like something just supercharged the oven, and instead of the muffins gently rising, they're suddenly starting to explode.

Isaac Newton proposed fundamental principles (laws) about the movement of objects that are still useful guidelines today. One such law was that an object in motion will stay in motion unless acted on by an outside force. Neglecting gravity and friction, if you throw a ball (say, in outer space) it will continue moving in a straight line at a constant speed. If our universe started with the force of a big bang, it would not be surprising that space is continuing to expand today.

But also according to Newton, for an object to *accelerate* (that is, to increase its speed) an additional force would be needed. If the universe is not only expanding, but is expanding at ever increasing speeds, then there should be a force that is causing that acceleration. Given the size of the universe, that is one heck of a force. Scientists do not know what the force is that is speeding up the expansion and have simply labeled it "dark energy."[10] Gravity also acts through the entire universe, but it has an attractive effect. Dark energy is a repulsive force, apparently much stronger than gravity.

Not understanding this force is a pretty big deal. At one point, physicists thought that the era of scientific discovery was over, and all that remained was fine-tuning the calculations in the laws of the universe. Now we've discovered the presence of a powerful force permeating the entire universe, and we don't know what it is. Equally mysterious is that scientists have discovered that there is a large amount of matter in the universe that we had not accounted for. It's appropriately called "dark matter."[11] It apparently is providing some of the gravity that holds galaxies together. Combined, dark matter and dark energy comprise about 95% of the essential stuff of the universe. Everything else that we can see or measure is just a small smattering of what's out there. There's a future in science after all.

Despite Einstein's speed limits for travel, space is allowed to expand at a rate even faster than the speed of light. Einstein concluded—and no one has disproved—that anything with mass must travel at less than the speed of light through space. But space itself is not traveling; it's expanding, growing from within. The net result is that some objects in space have moved further from some other objects in a time that would compute to a speed greater than the speed of light. It's like an earthquake opening a rift between two mountains: the mountains themselves aren't traveling, but the distance between them has grown.

This expansion has profound implications for our future, though nothing to immediately worry about. If the groups of stars and galaxies are moving away from us at speeds faster than light, we will never see them again. Their light will never reach us. The distance between them and us is growing faster than even light can keep up with. Billions and billions of years from now, there won't be much to look at. And even those few lights we can see will flicker out into a cold ash. Unless, of course, the whole shebang (as Timothy Ferris calls it)[12] suddenly stops expanding and starts contracting. Stranger things have happened.

How long has this expansion been going on? To varying degrees, the universe has been expanding for its entire existence. Making a few assumptions about this expansion, astronomers have traced the history of the universe backward in time. If there was a movie of the expansion, you could run it backwards and watch the contraction of the universe receding back in time.

We know a force that could cause that—gravity. In fact, until the accelerating expansion was discovered, it was thought that the moderate expansion of the universe would eventually slow down due to the overwhelming force of gravity resulting from all the stars and planets in the universe. The expansion would stop and the contraction would begin, resulting in a big crunch into a tiny ball.

Instead, the counteracting force of dark energy has been competing with gravity. Gravity's strength lessens significantly with increasing distance, so the force that is expanding the universe ultimately overcame the retarding effects of gravity and set it on a course of accelerated expansion about five billion years ago.[13]

The Beginning of A Time

Even though a universal contraction now seems less likely, it is still possible to trace the history of the universe back in time, culminating in its very beginning as an infinitesimal dense bundle of energy. Where that bundle came from is a far deeper question. But the best explanation of how we got to the universe we now have is that it began incredibly small[14] and rapidly expanded through what is now called the "Big Bang." It was neither big in size nor loud in its bang, but it was a big deal, so the name has stuck.

The path backward in time covers approximately 13.8 billion years, a very long time by any measure. However, in terms of the potential future life of the universe (which could even be infinite), it is just an instant. The Earth has been around for about 4.5 billion years, but scientists have theories about what the universe will be like hundreds of billions or even trillions of years from now.[15] From that perspective, 13.8 billion years is just a blink of the universe's eye. Why should we find ourselves as pioneers in such an early moment in the history of the universe?

Of course, this is the moment that a species (us) came into

consciousness of science, so it makes sense that awareness of the universe's age would take place now, if not before by another species on another planet. But from the universe's potential perspective, this is a very unusual time, assuming it goes on for hundreds of billions of years to come, with us or without us. The Big Bang Theory (not the TV show) inevitably raises many questions. What existed before the big bang? Was there any sense of time then, or was it endless before-ness? What caused the tiny bundle of energy to start expanding, and what was the force that powered it? For the big bang to produce the universe we now have, many cosmologists have concluded that there had to be a short period of enormous exponential growth that briefly occurred just after the initial expansion started. But what caused that? The particles and forces that predominate in our world are the subject of the second half of this book. But the very small and the very large intersect at the big bang. Unfortunately, our theories (like gravity) of how very large objects behave do not work well on the atomic scale, and our theories (like the Standard Model of particle physics) of how very small objects react have not incorporated gravity.

There are other difficulties in exploring the origins of the universe. For one thing, so much was happening in incredibly short periods of time. One of the first milestones in time after the initial big bang of the universe is known as the Planck epoch,[16] named after the German scientist Max Planck, who discovered that much of what happens in the world of the very small is not continuous but is broken into mysterious quantities—or quanta. Before the Planck epoch, the universe is a deep mystery. Particles did not exist, the laws of physics were yet to be formed, and even the dimensions of time and space were undefined. The Planck epoch lasted 10^{-43} seconds after the initial big bang. To simplify the expression of repeatedly multiplying a number (like 10) by itself, mathematicians use exponents: for example, 10^{43} is equivalent to 10 used as a multiplier 43 times. A negative exponent indicates 1 divided by 10 to the given power. Thus, 10^{-43} is 1 divided by 10^{43}, a very small number. We simply have no concept of that short a time. A tenth of a second is 10^{-1}. Races are sometimes decided by a tenth of a second. A hundredth of a second is 10^{-2}. Camera lenses are capable of opening and closing in that time. A millionth of a second is 10^{-6}. We don't have any common experience of things happening that fast, though light can travel about the length of a city block in that time. The Planck epoch is many, many times shorter than a millionth of a second. We're a long way from controlling experiments with such extreme time limits.

Such strange numbers can readily turn up in mathematical equations that are part of the theories of physics. But subjecting such units to experiment is a different matter. And yet, science is making progress exploring the universe at that incredibly early stage, and hoping to devise the laws of physics that applied at that time. Super atomic accelerators help provide the enormous energy, similar to what existed in a tiny space near the beginning of the universe. The Large Hadron Collider at the European Organization for Nuclear Research (CERN) in Switzerland, for example, accelerates two bundles of protons in opposite directions at nearly the speed of light around a 17-mile underground track and then slams the two streams into one another.

Such collisions simulate the concentrated energies in the early universe, though not simulating the big bang. Elementary particles emerge from these collisions, usually only lasting a fraction of a second before they decay into other more stable particles. Such experiments recently led to the discovery of the elusive Higgs boson (also referred to as the "God particle"), about which we will say more later.

Star Trek
As science reveals the mysteries of the universe on a grand scale,

it raises the question of whether we humans could explore this cosmos to its farthest reaches. At first blush, going into space appears to be only a question of developing better technology, something humans are very good at, given enough time and money. We've gone to the moon; we've sent robots to Mars; so, in the words of Buzz Lightyear, "To infinity and beyond!" Fortunately, it may be that space is not infinite, so we don't have to go that far. Unfortunately, much that is potentially interesting in space, like planets that are possibly similar to the Earth, are so far away that the space shuttle, traveling as fast as it has gone, would take millions of years to reach them. Obviously, the space shuttle is going to look like a horse and buggy in a few years. But speeding up our travel to planets orbiting other stars runs into more than engineering limitations. The universe has a strong resistance to any speed approaching the speed of light, and there is a firm law that says nothing can go faster through space than light. Even at that unattainable speed, it would take years to reach even some of the closest stars, and for most of the stars it would take millions or even billions of years.

If we can't beat light, perhaps at least we can go nearly as fast as light and visit some of the closest stars in a lifetime. Again, nature appears to be limiting us rather than charting the way to an unbounded future. According to Einstein[17] and well confirmed through experiment, something happens to us (and to all other objects) when we start increasing our speed: we get "heavier," as measured by stationary observers. As a matter of fact, the closer we get to the speed of light, the more resistance we incur to going faster because our mass is increasing, and approaches an infinite amount. (See Chapter 5.) Although science has constantly solved the seemingly impossible, the universe resists traveling at unbounded speeds.

There is one side benefit to faster travel that may provide a partial solution. Along with our mass getting greater, our clocks (including our internal aging clock) will slow down the faster we go, as measured by those standing still. On one level, we're familiar with time passing slowly or more quickly. When we're having a good time, time flies. Two hours at a great movie seems to go faster than two hours in a dentist's chair, but when we get back home we would see that our clocks advanced two hours in each case.

This is not what happens when traveling close to the speed of light. If I spend a few minutes (as measured by me) traveling at nearly the speed of light, when I land and get home again, my kitchen clock may have ticked two hours. Yet my wristwatch has only advanced a few minutes. This phenomenon is discussed more fully in Chapter 5 on Time and Space.

This can be very good if you have a long distance to travel. Once you get going at a fast enough speed, you will age more slowly because all of the aging processes will have slowed down, too. You may get back to Earth to find no friends left. They've all died in their own good time, while you have only advanced a few years. You've gone to the stars and back, but you will have to tell your story to a future generation.

In the chapters ahead, we'll explore other aspects of this strange universe in which we live. What are black holes and why should we care about them? What are gravitational waves and what do they tell us about the universe? Do other universes like ours exist, even if they're outside of our ability to detect? What are the implications of discovering intelligent life elsewhere in the universe? And finally, why is there something rather than nothing?

Chapter 2

Much More Than Stars

Nothing that is human is purely human and nothing that we see in the sky is purely cosmological. We are embroiled in the cosmos.
– Timothy Ferris[1]

Holes in Space

Black holes seem like an unlikely topic for a book on science that is accessible in ordinary language.[2] They are anomalies in the universe and unlikely to directly affect our daily lives. Although Einstein realized that his theory about gravity led to the possibility of black holes, he was very skeptical that they actually existed. Now they have joined our everyday conversation as a catchword for the mysterious and dangerous, things surely to be avoided at all costs.

One reason for trying to understand black holes is that they now appear to be pervasive throughout the universe. There is almost certainly one at the center of our Milky Way Galaxy, and at the center of most other galaxies as well. At first, astronomers proposed black holes as a special explanation for the bizarre behavior of a rare form of collapsing stars. Although gravity tends to contract all mass into small spaces, there are counteracting atomic forces that prevent a complete collapse, so black holes were thought to be very rare.

What Makes Black Holes So Unusual?

The most commonly mentioned and supposedly startling fact about a black hole stems from its name: nothing, *not even light*, can escape from the black hole's pull once it crosses an invisible boundary surrounding the hole, called the event horizon. But in a way, that should not be too surprising. If the path of light is bent by gravity, then it could get trapped by a heavy enough object, just as a satellite is trapped by the Earth's pull.

Light traveling through the universe will have its path bent towards any massive object, not just a black hole. If the light is close to the object, or the attracting object is heavier, light might even be brought into an orbit around the object. Finally, light might even be sucked into the object, unable to reflect out, just as a satellite may be drawn down and actually hit the planet. Some scientific theories would even allow for a "black star," that is, an object massive enough that it traps light but is not so heavy as to collapse to a single point.[3] If light is caught and cannot escape, there is nothing to see—just a void. Black is the absence of light.

Early philosophers believed that objects (like apples in a tree) fell to the Earth because they had a natural tendency to do so. What is up must come down. Isaac Newton introduced a new idea, radical at the time. He concluded that there is no up or down. And apples are like the moon, the planets, or any other object. All objects with mass (or equivalently "weight," here on Earth) attract other objects, with the force dependent on the size of the two attracting objects and the distance between them. Heavy objects exert a larger attraction on objects around them than light objects, but the pull falls off as the square of the distance (distance times distance) between them.

The moon is caught in the Earth's gravitational pull, but it, too, pulls on the Earth causing tides. The Earth is pulled towards the sun. We are also pulled by the gravitational effect of many other suns (stars), but they are far away, and when you square their distance it greatly diminishes the pull that the Earth experiences. Apples are attracted to the Earth, but they would also fall to the moon if it had any fruit trees.

But why should light be affected by the pull of gravity? In Newton's time, light was thought to be made of particles possessing a small amount of mass. Hence, light was pulled like any other weighty object. Newton's theory of gravity was modified by Einstein's theory of general relativity, which itself may be modified someday. Einstein and others concluded that light had zero mass, but he also had a very different notion of gravity.

Bending of Space

Einstein was uncomfortable with Newton's notion that two objects could instantaneously attract each other at great distances with no means of conveying that force. It seemed to violate the universal speed limit. In rough terms, he reintroduced gravity as a bending of space itself caused by the mass of any object. That bending of space alters the path of an object passing by, changing its path to be closer to the massive body. Sometimes that bending would cause the passing object to be caught in orbit around the massive body, and it might even bring it down to its surface, just as the Earth bends space, affecting the path of a rocket launched from the ground. The initial path of the rocket might be straight up (i.e., away from the Earth), but its path will be bent by the curvature of space, perhaps bent enough to cause it to fall back, or at least to stay in orbit. It all depends on the speed at which the rocket is launched compared to the massiveness of Earth.

If you are putting a golf ball on an uneven green, you can't aim it in a straight line to the hole and expect the ball to go in. You have to allow for the contour of the green, which is equivalent to your space for such a task. If the ground between you and the hole slopes to your left, you'd be wise to aim the ball to the right of the hole so that it will eventually curve left and fall in. If you're shooting an arrow and aiming it in a straight line at the target, you will miss. Newton says you will miss because the Earth pulls your arrow down before it reaches the target. Einstein says the space between you and the target is bent in the presence of the mass of the Earth. If you can picture the right amount of bending and aim a bit above the target, your arrow will travel the graceful arc of curved space to the bullseye.

In Einstein's world, light rays will be bent by massive objects not because light is attracted by the object, but because space is curved by the object and light travels through space. How distorted can that bending of space be? That is where the strangeness of black holes really comes in. Einstein did not just theorize that gravity is the bending of space, he actually provided the equations that could measure that bending. And by those equations, if the mass of an object is sufficiently compacted into a small enough volume, the bending produced tends to infinity.

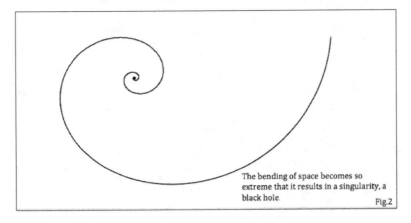

The bending of space becomes so extreme that it results in a singularity, a black hole.
Fig.2

That in a nutshell is the mystery of black holes. Infinity is a messy concept. An infinite number of objects will have an infinite mass and an infinite amount of gravity. In an infinite amount of time, anything that can happen will happen, infinitely often. Inside a black hole, the bending of space is so extreme that it forces the equations into infinite results. Further refinements are needed to fit these strange objects into a manageable theory. The capturing of light is just a small part of their enigma.

This does not necessarily make black holes the most fearful objects in the universe, capable of gobbling up everything over time.[4] A black hole formed from a massive star has no more gravity than the original star that gave it birth. (Black holes can acquire more mass and can even absorb other black holes, but their attractive force is still greatly diminished by distance and by the countervailing force of other objects in space.) A spaceship can safely stay at a respectable distance and observe a black hole, assuming it has enough fuel to escape its relatively minor pull. But drawing too close to the invisible boundary, its event horizon, will get you sucked in no matter how much fuel you have.

So, black holes are in many respects much like every other object in space. They have mass and their gravitational effect is the bending of space around them, distorting the path of anything in their neighborhood to a degree dependent on the closeness and speed of the other object. The strange thing is that the mass is all concentrated in a single point (called a singularity). The mass has no volume and infinite density.

The Making of a Black Hole

To appreciate black holes more, it might be helpful to have a general idea of how they form.[5] A typical black hole starts off as a star. Stars are objects larger than planets, but are much more volatile. Stars are massive enough that their primary element, hydrogen, is moving around with such energy that hydrogen atoms crash into other hydrogen atoms and fuse together forming helium. (The process is more complicated than a single step.)[6] The total mass of the end product of this reaction is *less* than the total mass of the hydrogen atoms that started it. Some mass is lost in the fusion process. Where does it go? It is converted into energy according to $E=mc^2$, where m is the mass lost, c is the speed of light, and E is the energy given off. Because c is such a large number, the energy released from the fusion of two hydrogen atoms is powerful, especially when it occurs billions of times per second.

Experiments are underway to extract the energy from the fusion of atoms on Earth in a controlled way, though so far it takes more energy to do so than can be harvested from the fusion reaction. Ordinary nuclear reactors use a different process that divides atoms rather than fusing them, a reaction that also converts some mass to energy, and this process can be better controlled.

Anyhow, the huge fusion reactor that is our sun and the other stars create an enormous outpouring of energy that keeps us warm and helps our crops to grow. It also keeps our star from collapsing into itself from the force of gravity. Eventually, however, the hydrogen fuel for this process will run out. For our sun, that might take another five billion years. The critical mass of colliding atoms will drop below a sustainable level, and the star's outward energy will be overcome by its inward gravitational pull. Stars like our sun will just stay at a colder state, not collapsing to a point because there are still internal forces that prevent atoms from coalescing too close together. But if the star is sufficiently massive (about twenty or more times as heavy as our sun), the gravitational pull will overcome the internal forces and the star will continue to collapse. How far does this process go?

Usually, when a theoretical graph of a physical process tends ever more slowly (asymptotically) down to zero (or up to infinity), there is a sense that, in practice, it will never get there. In a collapsing star, gravity is bending space to such an extent that it curves back on itself. Its volume keeps shrinking until it occupies no space at all. That is indeed a mysterious concept. How does something occupy no space? If it doesn't quite shrink to zero, what is the powerful force that prevents it?

Not all black holes are collapsed stars. Some of the black holes at the center of galaxies appear to be billions of times more massive than our sun. It is unlikely that a star of such magnitude ever formed, and so astronomers are not sure how supermassive black holes emerged. Clearly, once a black hole exists, it can continue swallowing more matter around it, and thus grow in size.

Apparently, supermassive black holes began forming relatively soon after the big bang. Astronomers recently recorded one of the oldest (farthest away) such entities at a distance of 13 billion light-years away. That means it was formed about 800 million years after the big bang, or when the universe was only 6 percent of its current age.[7]

Black holes appear to hold other mysteries. If we sent a copy of every book ever written in a rocket to a planet in the far reaches of space, we might never get those books back, but the information in them would not be destroyed. If found by another life form that could translate our languages, all the information could be obtained. But if that rocket got sucked in by a black hole, all that information seems to disappear. Does it get scrambled down into an infinitesimal point from which it can never be reassembled or recaptured? Is information irretrievably destroyed in a black hole? The late physicist Stephen Hawking, among others, thought that the information might leak out over trillions of years as the black hole disintegrates. You can read about it in *The Black Hole War* by Leonard Susskind.[8]

Einstein's theory of relativity also implies that gravity (or the presence of mass) slows time down from the perspective of a more distant observer, just as time slows when traveling at a faster speed (see Chapter 1). In a black hole, time stops completely. That is, if you were able to peer inside a black hole into which a clock was falling, that clock would tick more slowly as it approached the singularity point, and would eventually stop altogether—not because the clock would be destroyed (which it would be), but because every kind of timekeeping device would be slowed to a halt: the beating of a heart, the vibrations of a molecule, etc. Stopping time is a notion foreign to our experience, just as is trying to understand whatever "came before" the big bang, or after the universe ceases any movement whatsoever.

Black Hole Time Machine

A black hole is truly an anomaly in the midst of an otherwise frenetic universe, which constantly vibrates with real and virtual particles and radiates waves of energy. A black hole could be a pathway to a new part of the universe, where the usual rules don't apply.[9] Some scientists have speculated that under the right circumstances, one black hole could align with another in such a way that a shortcut through space (called a wormhole) could be created. Think of a worm taking a shortcut through your apple rather than traveling around the outside. Instead of traveling thousands of light-years through normal space and time, you might be able to traverse a wormhole and pass directly to another distant location and arrive at a very different time.

Another way to think of a wormhole is to picture a winding path up a mountain, with switchbacks left and right so as to avoid a very steep ascent. Switchbacks slow you down because you have to travel a greater distance, but at least they make each step more manageable than going straight up. The more energetic hiker might decide to save time and jump directly from a switchback heading in one direction to a higher point heading in the other direction by clambering straight up the hillside. That's a rough analogy of using a wormhole for space travel. Although it's doubtful that humans could ever benefit from this, it certainly points to the strangeness associated with a black hole.

Gravitational Waves: Ripples in the Sea of Space

Another interesting topic whose implications for practical use are a lot closer than opening wormholes is the recent confirmation of gravitational waves in 2015. They, too, stem from Einstein's theories of relativity, and their recent detection in Louisiana is related to black holes. As mentioned above, gravitational attraction is not the result of some instantaneously transmitted force that emanates from every massive object. Rather, Einstein described gravitation as the result of the bending of space itself, and that all objects are immersed in a gravitational field that fills space.

A field is like a grid over a particular area. Every point of a field is associated with a particular value. The oceans are like fields of water with various heights and movement at every location. Energy can be transmitted across the field of water through waves that travel and crash on the shore. Fields in space are more complex, but the basic idea is a common one.[10] Football is played on a field, with yard markers measuring the distance to the goal at any point.

Light travels as a wave through an electromagnetic field that pervades all of space. We're familiar with waves in the water. Sound also travels as waves through the air. When the wave of air hits our eardrum, it causes the drum to vibrate in a way that can be interpreted by our brain as a sound.

If you want to feel what it's like to generate a wave, pick up piece of rope. Move one end rapidly up and down, and you will see a wave forming along the rope and moving away from you. The wave travels horizontally, while your movement of the end of the rope is vertical. The greater your up-and-down distance, the higher the peak (amplitude) along the rope will be. If you increase the energy (frequency) of your up-and-down motion, more highs and lows will be created along the rope. The field that the rope wave travels through is the rope itself. If you tie the other end of the rope to a wall, the wave you create will be reflected back towards you.

The light that we see is the visible part of a variety of waves that travel through the electromagnetic field, which can also contain radio waves, microwaves, X-rays, etc., depending on the frequency of the wave. Communications are said to travel through the "air waves," but that is a misnomer, since no air is moved and none is even required for TV, radio, or any such communication. All waves in the electromagnetic field travel at the speed of light—basically, they are light, just not all at the frequencies that we can see.

When science speaks of the speed of a wave, it is talking about the distance between two consecutive crests of the wave (its wavelength) per unit of time. The wavelength of a wave is the distance between the high point of one wave crest to the high point of the following wave. For water waves, the wavelength might be a few feet on a windy day, and the time between crests would be equivalent to the rhythm of the waves crashing on the shore. The wavelength of some radio waves can similarly be a few feet, but their speed is a lot greater.

The frequency of a wave is the number of crests that occur per second. High frequency waves like light have a lot of crests per second; hence their wavelength is relatively short. Radio waves have lower frequencies. Sound, too, can have many frequencies, only some of which are detectable by the human ear.

Did You Feel That?

Einstein proposed that a disturbance of gravity (the bending of space) also travels through space at the speed of light via the gravitational field. The ups and downs (it's actually three-dimensional) of the field are due to the passing of a gravitational wave emerging from an excitation in space caused by the movement of massive objects. For example, if two huge stars were to collide, that disturbance would result in a dispersal of gravitational energy that travels through space like a sound wave through air or a water wave across the ocean. Gravitational waves might also be called space waves, because space is being waved, just like water waves move water up and down.

We may think of water waves as moving water horizontally, say east to west. But if you watch a boat that's anchored to a spot as the waves go by it rises and falls in a regular pattern. If you've ever tried to get into or out of a floating boat being lifted by waves, you know how unstable that can make you feel. There is that moment when you have one foot in the boat and you have to lift the second foot off the land, but the boat is bobbing up and down. If you have nothing to hold onto, you can end up pretty wet.

Because a measurable gravitational wave typically emanates from a great distance away, we do not feel much of a disturbance when it passes through the Earth. If someone shouts your name in California you will not hear it in New York even though the shout produced a sound wave that travels and spreads out around the globe. When it reaches New York, its energy is so dissipated that it is imperceptible. Even a tsunami on one continent might only be felt as a gentle wave on a distant shore.

So, too, with gravitational waves. Einstein thought they would never be detected, just as he thought black holes would always remain theoretical. But gravitational waves were recently

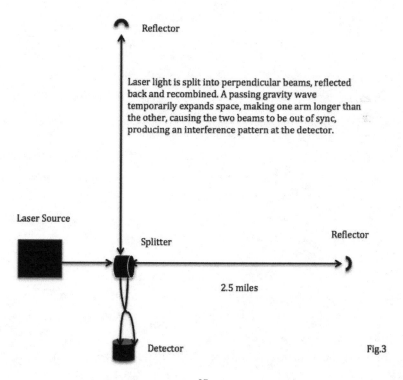

Reflector

Laser light is split into perpendicular beams, reflected back and recombined. A passing gravity wave temporarily expands space, making one arm longer than the other, causing the two beams to be out of sync, producing an interference pattern at the detector.

Laser Source

Splitter

Reflector

2.5 miles

Detector

Fig.3

detected by two independent experiments in Louisiana and Washington. Like the water waves that made getting into a floating boat difficult, the gravitational waves actually distorted space and then set it right again. It was a ripple in spacetime.

How was it detected? Scientists split a beam of laser light and sent it down two long perpendicular paths, shielded from other influences.[11] The light traveled down each path, and it then was reflected back by mirrors at the end of each path. The two beams were then sent to a single detector at the end of the final path. If a gravitational wave created a disturbance in space, that would cause the distance between the light source and the mirror to be slightly increased in one arm and decreased in the other (perpendicular) arm so the two beams would travel different distances than if they had proceeded without the ripple in space.

If both parts of the beam had traveled the exact same distance, then they would completely cancel (or completely reinforce) each other at the detector. That is, the crest of one and valley of the other would line up exactly making for a zero signal. But if one beam traveled a slightly different distance than the other, the beams would arrive at the detector slightly out of sync. This causes an interference pattern of light and dark bands that is detectable with sensitive instruments.

Two sound waves can similarly interfere with each other. With the right adjustments, the two interfering waves will cancel each other out completely, which is the secret behind noise-reduction earphones that some people wear on airplanes.

After the instruments were properly set up and refined, the facility in Louisiana detected such an interference in its light waves, indicating a gravitational thunderclap had occurred a billion light-years away. As a check on the whole process, an identical detector in Washington also registered the same interference at a slightly different moment, since the gravitational wave passed over the two locations at slightly different times. The fact that the disturbance was detected in two places independently and conformed to the predictions for how strong the wave emanating from the collision of two massive black holes would be confirmed the existence of gravitational waves (and earned the scientists a Nobel Prize).

Although this is a groundbreaking discovery and an example of physics at its highest level, it also demonstrates how a relatively simple and understandable apparatus can lead to an amazing insight into how the universe operates. The basic idea of an interferometer for the Laser Interferometer Gravitational-Wave Observatory (LIGO) in Louisiana was used to experiment with light by Albert Michelson and Edward Morley in 1887, though they were not looking for gravitational waves and they did not have lasers.

But before you try to detect gravitational waves at home, you should realize that the difference in distance traveled by the two light beams as a result of the gravitational wave was many times less than the width of a proton. This is an extreme degree of precision, and only the wave properties of light and its unique speed, along with the meticulous work of the scientists and engineers involved, made this possible.

There are now four such detectors around the world, and other black hole collisions have been recorded. Most recently, a different kind of collision was also detected. Measurements in 2017 indicated a gravitational wave resulted from the collision of two neutron stars.[12] Neutron stars are stars that have exploded as their fuel was winding down. After the explosion, they shrunk in volume, just as some stars do on their way to becoming a black hole. But with a smaller amount of mass pulling things together, some collapsing stars do not proceed all the way to the singularity of a black hole. Nevertheless, this is a massive collision, felt throughout the universe, or at least at those lucky places like Earth where sufficiently sensitive instruments have been set up.

Although not of immediate consequence in our daily lives,

detecting black hole collisions gives astronomers a new way to study the changes in the universe that are undetectable through telescopes. It's hard to see the union of two black holes since they don't emit light. (On the periphery of the black hole, there can be considerable detectable radiation as mass is being violently sucked in and dismembered as it crosses the event horizon.) In the case of two neutron stars, it can be a signal to astronomers to turn their telescopes to a particular point in space where an immense explosion is now taking place. (Actually, it took place 130 million years ago, but the light and the gravitational wave are just reaching us now.)

One of the most powerful rumbles in space occurred at the dawn of our universe, sending out gravitational waves almost 14 billion years ago. It is hypothesized that the effect of those waves will be imprinted on the faint glow of background radiation that permeates space as a result of the big bang.[13] Scientists in Antarctica are trying to pinpoint that effect by studying the background radiation.

Chapter 3

One Universe or Many?

Once you've discovered it's easy to make a universe out of an ounce of vacuum, why not make a bunch of them?

– Craig Hogan[1]

The Observable Universe

The universe we know is the one we can see, either directly through observation of the stars, or indirectly by observing evidence left from the enormous expansion known as the Big Bang. Einstein formalized the idea that time and space are both measurements within one entity called spacetime, which thus has four dimensions: the three common ones of space and one of time. That time should be intricately related to space makes sense. Whenever we observe something at a distance, be it a faraway star or a person across the street, we are really seeing that object as it was some time ago because the light that enables us to see the object takes time to reach us.

As the world's most powerful telescopes scan the night sky, they can detect light from stars that were formed shortly after the universe was born. Astronomers believe our universe is 13.8 billion years old, and that the first stars were likely formed shortly after its birth, about 100 million years later. That may not seem like a "short" time, but it is less than 1% of the current age of the universe.

The furthest objects observed so far are galaxies that appear to be just emerging, estimated to be about 13.4 billion light-years away.[2] Given the connection between time and distance, we could also say that astronomers have observed the presence of stars and emerging galaxies in the universe as it existed 13.4 billion years ago. The stars in those early galaxies might have since exploded long ago, but we'll have to wait for the light from those subsequent events to reach us to see such changes.

The present breadth of the universe of which we have had an opportunity to observe is much larger than 13.8 billion light-years, the distance that a photon of light released in the big bang has traveled. For one thing, when the universe began, it expanded in all directions. So there would be 13.8 light-years of space in one direction, and an equal amount in the opposite direction. But even that doesn't account for the fact that the universe is expanding. The universe is not just getting bigger, its growth *rate* is also increasing. For at least seven billion years, the expansion has been constantly accelerating, setting some objects apart at faster than the speed of light. So, for example, if, over one year, space expanded by one light-year of distance in every direction, the next year it will expand by more than a light-year. The net result of this growth is that the width (or the greatest distance between two points in the universe) is estimated to be 93 billion light-years.[3]

That's quite a neighborhood to be living in. It would take a beam of light—the fastest traveler in the universe—a billion years to travel about 1% of the distance across the universe, as it exists now. So far, all of the life we know of in this vast expanse sits on one planet near one star in a galaxy of 200 billion stars, which in turn sits in a universe of hundreds of billions of galaxies. The fraction of the universe that we have explored is infinitesimal. Given the number of opportunities for an Earth 2, the odds seem pretty good that we will eventually discover new neighbors.

Inflation

What we have been able to observe of this universe has raised some important questions about the theories describing how it all came to be. Our universe is fairly uniform anywhere you look, even though there are certainly differences when you get down to the finer detail, such as the Earth. Groups of galaxies called superclusters appear in every direction. Space also has a relatively even temperature wherever it's measured, a couple degrees above absolute zero, or about 455 degrees below zero Fahrenheit. A heat-map drawn from satellite observations of the radiation produced in the big bang—the Cosmic Microwave Background (CMB) radiation filling space—has variations of less than a degree.

In physics, the temperature of an object is a measure of the movement (average kinetic energy) of the particles making up the object. The temperature at which all vibration stops is called absolute zero on the Kelvin scale, whose degree units are equal to those on the Celsius scale, but with a different zero point. The uniformity of the temperature of the radiation filling space implies that there was some interaction of the particles in the universe just after the big bang that helped homogenize the eventual result. The problem for cosmologists is that the initial rate of expansion after the big bang would not have allowed for such mixing.[4] By the time particles coalesced into being from within the high-energy soup and light began escaping, they would have been too far apart to interact into a more uniform mixture.

To account for the world we find ourselves in, an ingenious but complicated theory of the universe's first moments has been proposed to allow for greater interaction and a more homogeneous mix of elementary particles. The theory is called Inflation and was first proposed around 1980, principally by Alan Guth and Andrei Linde. It hypothesizes that there must have been an enormous burst of expansion energy just after the big bang causing the tiny universe to grow exponentially for a short time, while retaining the closeness of the elementary particles at the early stages of the inflation.[5] The force was the opposite of gravity, expanding instead of contracting.

It might seem counterintuitive that speeding up the expansion of the universe would allow *more* time for the elementary particles to interact. However, if the growth process is sped up during one period that means it could be slowed down at a prior period and still result in the cosmos we have today. If you could get to your grandmother's house in an hour by driving at a constant speed, you could also arrive in an hour by going faster than that speed during one part of the trip and slower at an earlier portion. The slower growth rate of the universe at first could have allowed for the particle interaction that eventually produced the relative uniformity we see today.

The numbers involved in Inflation are extraordinary. The theory proposes that the sudden burst of expansion occurred about one trillionth of one trillionth of one trillionth (10^{-36}) of a second after the initial big bang. No current technology would come even close to detecting such a short period of time. The super expansion itself lasted a tad longer, about a billionth of a trillionth of a trillionth of a second, pushing the universe to the size of a grapefruit. That may not sound like such an impressive growth except if you realize that prior to this time the universe was much smaller than the size of a proton. This is an expansion rate of a hundred trillion trillion times (10^{26}). The expansion of the universe then settled down to a more moderate pace.[6]

Of course, a completely homogeneous universe, with a uniform distribution of particles everywhere, would not be very interesting—no stars, no planets, no us. Fortunately, the tiny irregularities in the early mix caused by the inherent fluctuations of subatomic particles eventually became the seeds for molecules, stars, galaxies and all the other fascinating structures of the universe.

Inflation theory has been widely accepted among scientists because of the difficult problems it helps solve.[7] Measurements of the Cosmic Microwave Background (CMB) radiation mentioned above reinforce the theory of Inflation. This radiation has a uniform appearance and temperature, just what Inflation would predict.

Although the CMB had been theoretically proposed earlier, it was discovered accidentally by two scientists at Bell Labs working in New Jersey in 1964.[8] They detected an annoying hiss from an antenna they intended to use for satellite communications. When their best sleuthing could not uncover a conventional source for this electromagnetic signal, which was heard from every direction that they pointed their antenna, they asked for help from physicists at Princeton. Those scientists realized the signal might be coming from the photons filling all of space shortly after the big bang and now dispersed throughout the universe. The finding was reported to the scientific community, and the accidental discoverers were awarded a Nobel Prize.

Until fairly recently, people commonly saw evidence of this CMB radiation on their TV sets. If you had a tube TV with an antenna for reception, the fuzzy "snow" that showed up between regular channels was partly caused by the CMB signal from space. Digital TVs and cable reception have eliminated that view, but the radiation is still out there.

Further evidence supporting the Inflation theory appeared to come recently from the purported effect of gravitational waves on this background radiation, as reported from a research station in Antarctica. That breakthrough, however, proved to be premature because it failed to allow for the fact that cosmic dust might have imitated the effect of the gravitational waves on the radiation, rendering the conclusion unreliable.[9] This errant detection of gravitational waves is different from those of the later detections at the LIGO observatories mentioned above (Chapter 2).

Despite the usefulness of the Inflation theory, there are rumblings within the scientific community about its validity. It requires a very unusual injection of expansive energy into the nascent universe at a very precise time. Although theoretical explanations have been offered for the source and timing of the inflationary push, it has the feel of a fudge factor to some experts.

It does answer questions if it really happened, but determining what was occurring in a tiny fraction of a trillionth of a second after the big bang is largely hypothetical.

The Multiverse

Another problem with inflation, although not a contradiction, is the conclusion by the originators of the theory that if inflation did trigger an early rapid expansion of our universe, it is inevitable that a similar inflation will be occurring in other parts of the universe and there will be an infinite chain of such explosive expansions.[10] Inflation is associated with a unique field (the inflaton), and that field can take different values in different areas of space. Some of those values will result in new rapid expansions. In other areas, such as that occupied by the portion of the universe that we can observe, the value of the inflaton dropped back down so that only ordinary expansion occurs. Each of the expansion pockets creates a new "universe," comparable to our own, but forever separate from ours and likely with different physical laws. The totality of all of these worlds is called the multiverse.

An endless chain of new universes popping into existence but remaining undetectable is the kind of theory scientists dislike. If it can't be tested, it's more like a philosophical conjecture. Science deals with the universe we live in and with things that can be measured in some way. It could still be that ways of confirming inflation will be found or that interactions with other universes (say, the collision of two such worlds) will be detected, but in the meantime alternatives to Inflation are being explored.

Although our universe is expanding and the rate of expansion is speeding up, there is still the chance that the dark energy behind that expansion will dissipate and be overtaken by the countervailing influence of gravity. The universe could start shrinking when that happens, running the movie of expansion in reverse, though time will not be going backwards. If gravity prevails, our universe could end in a big crunch. This could then be followed by another big bang. Moreover, our big bang— nearly 14 billion years ago—may have been preceded by a big crunch of the previous version of the world. And so on.

This continuous cycle of universe-creation allows our present universe to have the homogeneity it exhibits without injecting an unusual shot of inflation at a very precise moment. And instead of wondering why and how our universe started when it did, we're left with a continuous cycle of expansions and contractions. In a partial answer to the question of why there is something rather than nothing, we're left with the conjecture that there has always been something. That may be no easier to explain than one expansion, but there is no escaping some strangeness in theorizing about the universe.

Other Theories of the Multiverse

Other hypotheses also involve a universe beyond the one we're just beginning to explore. The simplest form of a multiverse stems from the fact that the universe we see is a function of both the power of our telescopes and how far we can look back into time. The earliest years of our universe contained no stars or visible matter. The particle soup was too hot to allow any coalescing of matter or star formation. When the first stars formed some 100 million years after the big bang their light began a long journey. Thirteen and a half billion years later that light is approaching the Earth and that is the limit of the earliest stars we can see. But if other stars exist even further away from us than the distance covered by light since the universe's inception, then those stars are currently beyond our view. They are outside of our visible universe.

How could stars have gotten so far away when their speed is bounded by the speed of light? Recall that the stars escaping our view are not so much traveling through space as being separated from us by the expansion of space itself. They are the blueberries in the rising muffins, gradually growing farther apart. The expansion of the batter of space can separate objects at an equivalent speed greater than that of light.

So, there may be a large part of the universe that is beyond our view and exploration. It probably looks like the universe we can see and follows the same laws if it had the same beginning. But we may never know for sure.

Could the extent of this universe (the observable plus the unobservable parts) be infinite? Yes, though any infinite amount of physical objects creates some challenges. Similar to what happens in an infinite amount of time, in an infinite spatial universe, every conceivable world can be found and will occur an infinite number of times. The possibilities, as they say, are endless, including every conceivable variation of our world as well as exact copies of our world, and us.

Of course, you might rightly ask, "How can the universe *not* be infinite?" If you go over the edge of a finite universe, what would you fall into? Is there a wall at the border? That seems logical from the point of view of Euclidean geometry, the kind most of us were taught in high school. Parallel lines never meet and the angles of a triangle always add up to 180 degrees. But although it goes beyond the scope of this humble book, there are other geometries that can fully describe a universe.[11] If you try to use Euclidean geometry to plan a sea voyage, you will be in trouble. On the surface of a globe, every path is curved and the angles of a triangle do not add up to 180 degrees.

It's quite possible for a region to be both finite and yet unbounded by edges or walls. You can travel as long as you like on a globe and never hit an edge. And when it expands — like the surface of a balloon being blown up — objects on the surface become farther apart even though they haven't traveled on their own. This discussion gets a little tricky when we think of the three spatial dimensions of our universe. The surface of a globe or the side of a piece of paper is two-dimensional. (The globe itself, like the solid Earth, is three-dimensional, but the surface can be navigated with just two: latitude and longitude.) Although it's hard (perhaps impossible) to visualize the three-dimensional equivalent of the surface of a sphere, there is no reason it can't exist. It could be the "surface" of a four-dimensional sphere, just as the two-dimensional surface of the Earth envelops a three-dimensional globe.

A circle consists of all the points in a two-dimensional plane equidistant from its center. The circle itself is just one-dimensional. The surface of a sphere consists of all the points in (three-dimensional) space equidistant from its center. The surface itself is two-dimensional. To get to a three-dimensional curved space, consider all the points in a four-dimensional space equidistant from a given center.

We tend to rebel against a four-dimensional world, just as the creatures in the book *Flatland* by Edwin Abbott[12] rebelled against a world with more than two dimensions. Mathematicians and scientists, however, have imagined space with many more spatial dimensions than three. Just because we haven't experienced another dimension doesn't mean it doesn't exist. An ant walking along a thin wire is restricted to one dimension — back and forth along a line. But if the wire were thicker, the ant might discover that it can walk around the wire, as well as back and forth. It just doesn't notice the second dimension when the wire is so thin.[13]

All this discussion is only to make the point that there may be more to our universe than we can ever observe. More exotic versions of the multiverse emerge from other theories. One proposal has a new universe formed every time some event could have more than one outcome. Instead of choosing one outcome, the many-universes explanation proposes that both outcomes occur, each in a similar but separate universe.[14] It's like if it might be rainy or sunny on a particular day, there will be a universe for each, with you in them experiencing both forecasts. Of course, the you who was consciously facing the prospect of either rain or sun will end up having one or the other. You'll never know how the you in the rain that day turned out. Just enjoy the sun.

This theory is actually more sophisticated and involves the mysterious quantum nature of reality that we will discuss when we explore the universe on its smallest levels in the second half of the book. The short form of this idea stems from the evidence that the fundamental particles that make up everything we see in the universe do not occupy a simple location at a given time. Their location is spread out. But when one of those nebulous particles is observed or measured, it suddenly reveals itself in just one place. How that happens is unknown, and some scientists believe it doesn't really happen. Rather, they say, the universe branches into many paths when a particle is observed. All the paths are taken, though as one observer you just see one path. The other paths are in other universes with other yous doing the observing. This creates a multiverse with a huge number of universes, but at least it avoids the question of why the act of observation should so radically alter a particle.

String theory, which is touched on in the last chapter of this book, is a complex attempt to explain all the forces and particles in the universe. It posits tiny one-dimensional strings as the fundamental entity of the universe from which everything is built. Recent variations of the theory indicate the strings could produce 10^{500} different universes, so the number of possible alternative universes seems limitless.

Most versions of the universe could be dead ends as far as life is concerned. How is it that we ended up in one of those that allowed life, and on a planet where it could thrive? The simple, but to many unsatisfactory, answer is that we wouldn't be asking these questions unless we were so fortunately situated. The history of science is filled with surprises and the overturning of common wisdom. Just because the Earth looks flat, or it seems we are the center around which everything revolves, doesn't mean it's true. We may only experience one universe, but that doesn't mean there aren't many others. If one can exist, why not many?

Chapter 4

Life on Other Planets

Every atom of iron in our blood was produced in a star that blew up about 10 billion years ago.

– Jill Tarter[1]

One of the reasons humans have always gazed with wonder into the night sky is to ponder whether we are alone in the universe. Perhaps someday we will be visited by intelligent life from somewhere else. Or we might need to leave Earth for more room or a new home entirely. Where would be the best place to go? Is there anywhere else that would be hospitable to our species?

Finding life in another part of the universe could be very revealing. We might find advanced civilizations that help us answer questions that have stumped us for millennia. How and why did the universe begin? Is it possible to go backward in time? Can anything travel faster than the speed of light? Can life be endless? Many questions.

Contact

Of course, it is not guaranteed that other intelligent life would be friendly. Our science fiction stories are full of space invaders intending to do us harm. We react violently and somehow always prevail, despite overwhelming odds. Chances are, if we discovered intelligent life on another planet, we'd prepare to do battle.

Given that prospect, why don't we just keep quiet? We don't need another planet to live on right now, and it seems there are places we could live within our planetary neighborhood that are almost certainly devoid of potential competitors. The resources needed to live on a place like the moon or Mars would be significant but not insurmountable.

If we send out signals or physical probes, they might reveal to other species that we exist and where precisely we can be found. In our solar system, there is no other lush green body with ample water and oxygen besides Earth, should aliens be in the market for that sort of thing. Somehow, we can't resist saying, "Is anybody home?"

In an indirect way, we have been sending messages to potential intelligent life in the universe for decades. Ever since we discovered that information can be transmitted through space by creating waves in the electromagnetic field that permeates the whole universe, we have been broadcasting radio, television and Internet signals across our planet. Those waves also travel forever into space, although their intensity diminishes as they spread out, just as the gravitational waves from colliding black holes caused just a tiny ripple in our local spacetime in 2015.

It is hard to know what alien species might think of programs like *I Love Lucy* or *M*A*S*H*. In the right context, they say a lot about our self-conscious humor as an intelligent life form, but taken in isolation, they might create a different picture. Unfortunately, one of the earliest television broadcasts was Adolf Hitler's speech at the opening of the 1936 Olympics in Berlin.[2] In any case, there is an abundance of information about us and our planet floating through the universe for anything capable of listening and interpreting what is going on.

How far have our signals gone? Radio waves have been transmitted for a little over 100 years, television signals have been sent for over 70 years. Since electromagnetic waves (light, radio, TV, etc.) travel at the speed of light, those radio signals have traveled about 100 light-years. That's plenty far enough to reach many stars and their accompanying planets, some of which are less than ten light-years away from us. But it's a long way to the next galaxy, which is about 70,000 light-years away. So relative to the entire universe, our signals have not gone very far, and they are degrading significantly as they continue to spread. They were, of course, never intended for interstellar communication.

One of our more deliberate attempts to communicate with the universe was placed aboard the Voyager 1 satellite, which was launched in 1977, and which is just now leaving our solar system.[3] Traveling much slower than the speed of light, it has covered about 13 billion miles and still communicates with Earth. Thanks to the work of Carl Sagan and others, a gold disk with etchings of information about Earth was included in this early attempt to explore the outer planets in our solar system. (A similar disk was placed aboard Voyager 2, which was launched around the same time.)

The disk contains greetings in 55 languages, messages from the Secretary General of the United Nations and the President of the United States, along with a wide variety of music styles, from Chuck Berry to Mozart. There is also a drawing of the human form. In the unlikely event this satellite ever gets retrieved and the disk read by another intelligent species, they may have many questions, but they will at least gather that some other intelligent life exists in the universe and is reaching out beyond itself.

The most recent example of human culture launched into space was aboard the Falcon Heavy test rocket developed by Elon Musk.[4] The test rocket with three boosters was topped with a capsule containing a cherry-red Tesla Roadster and a manikin driver. The launch on February 6, 2018 was so successful that the payload was sent far into the asteroid belt. The driver, Starman, however, may serve more as an inspiration for Earthlings than a message to aliens.

Probability of Other Life

What are the chances that our search for other life will be successful? From a scientific perspective, this is a question of whether the raw ingredients of life exist in other places and how many extraterrestrial environments are there that could sustain such life. Beyond the realm of science, there is the possibility that the human species is indeed the purpose of all creation and the rest of the universe was made to complement us. In that scenario, other intelligent life is less likely, and we have lots of room to roam. Of course, it is also possible the creator set in motion many forms of intelligent life throughout the universe. Earth-based theologies might need only a minor tweaking to include such beings in their narrative. Perhaps we and other intelligent species can all get along, as we sometimes strive to do on Earth among our different peoples and among our "aliens" of plants and animals.

Setting theology aside, life elsewhere in the universe seems very likely indeed. The universe is full of stars, though we do not expect to find life there. Life as we know it is impossible in a gaseous atmosphere with a temperature of 10 thousand degrees resulting from constant nuclear explosions.

We have been to the moon. Not much there, but there is probably some water below the surface, and where there is water on Earth, there is usually life. Mars is another potential location for a form of life. We have explored this nearby planetary neighbor of ours with many satellites, landings, and two successful rovers. The pictures have not revealed creatures cavorting across the plains, or even a tree or bush. There is evidence that abundant water flowed on Mars at one time and may lay below the surface even now.[5]

If we land astronauts on Mars, which seems likely over the next 20–30 years, they can test for life in the most probable places. Advanced rovers could also accomplish that task. It could be that we will find some fossilized form of past life. Perhaps single-celled creatures evolved when conditions were different, but died off when the climate changed, choking off the path to further evolution.

Finding any life, either past or present, might answer one

critical question: can life be based on a different plan than the one exhibited by all species on Earth? All Earthly life points back to common ancestors because it all contains DNA in its cells. The code for building each species exists in each of that species' cells. The code is partly different for bananas and humans, but a lot of it is the same, and all of it uses the same form of chemical instructions.

Perhaps life on Mars had a completely different way of sustaining itself, using different ingredients than the chemicals used here. If so, that opens more possibilities for life elsewhere because the same ingredients as used on Earth do not have to be present. Even if life is not found on Mars, we might decide to start a human colony there, partly because we love to explore and inhabit new lands, and partly because we may need a backup to Earth.

Venus is also relatively close to us, but its atmosphere would be deadly to humans, and even navigating through its shroud would be very difficult. The gaseous giants further out in our solar system, such as Jupiter, Saturn, Uranus, and Neptune, are not the most likely home for life, at least as we know it. The moons of some of those planets, however, bear some interesting characteristics.

Enceladus is a moon of Saturn. The Cassini probe of Saturn, which recently concluded its mission by descending into the ringed planet, also closely examined Enceladus. It passed through what appears to be towering eruptions of steam and water, reminiscent of the geysers in Yellowstone, but spouting much higher.[6] That tells us two things: there is water under the surface of this moon, and there is energy to heat that water and send it miles into space. Those are two vital ingredients for life on Earth. Moreover, in the geysers there are chemical signs of underwater formations that on Earth exist in the deepest parts of our ocean, where sunlight never reaches. These underwater spouts in our seas are warm enough and nutrient enriched, harboring a vibrant living community. Indeed, there is strong evidence that all life on Earth began in such challenging environs. If there are hot vents under the crust of Enceladus with other chemicals that elementary life can feed on, then the evolutionary chain of species may have a foothold there. The moon is not very large, about 500 miles in diameter. Its surface is icy white and cold. But below the surface might be a thriving colony of elementary life forms. The Cassini project has shown we know how to get close to Saturn and its moons. The rest (like actually landing on Enceladus) is just a matter of fine-tuning the technology and committing the resources to such a mission. Of course, such a journey would have to be very careful not to contaminate any Enceladus life with our own life forms.

Other moons in our solar system may also harbor life. Titan, another much larger moon of Saturn, has significant quantities of methane, which is a harbinger of early life. Io is a moon of Jupiter and may be similar to Enceladus in that water is likely present in some form there, too. Io is much larger than Enceladus, making it closer in size to the Earth. If life exists there, it may be nearer the surface. We're not likely to find species like ourselves on either of these moons, but the next candidates for likely habitats require traveling to another star, a journey of much longer duration.

Returning to the question of intelligent life in the universe, the number of possible locations for such life has grown enormously as our exploration of space becomes more expansive. As mentioned before, it is now apparent that we are just one planet in one solar system among over a hundred billion such suns in billions of galaxies, many of which are larger than our own. (To say nothing of other universes, which we may never reach or even be sure of their existence.)

Stars are not the abode for life, but they are the key ingredient to provide the energy that keeps planets in a relatively steady state so that life can grow, and many of the stars can provide life-sustaining warmth to planets for billions of years. On Earth, that lengthy time period has allowed a blossoming of life in almost endless forms.

Our star has eight planets, since Pluto was demoted in 2006 to a lesser status: "dwarf planet." If our solar system is typical, the number of possible planets to cradle life might be many times the number of stars—billions and billions and billions! (Not counting the moons of such planets, some of which are planet sized. Saturn, for example, has 62 known moons.)

Not every star is in a position to benefit its planets in the way our sun does for Earth. Our sun sustains life on Earth because we are in the "Goldilocks zone" of the sun, that is, not too close and not too far, but just right for warmth. Any closer and the sun might boil all our liquid away; further away, everything might be frozen in place. Other planets in our solar system are not as fortunate.

Another thing about stars is that they don't last forever. The larger ones burn out more quickly, and many explode, with their remnants collapsing into unfriendly bodies such as neutron stars, black holes, or black dwarfs. Their planets suffer accordingly.

Our sun, fortunately, is right in the midst of its projected ten-billion-year life span. The Earth was formed about 4.5 billion years ago, shortly after our sun was born. So, life has had a long period to develop into existence and evolve into the many and wonderful forms it takes today.

Does the emergence of life occur wherever the right chemicals, conditions and duration are present? We don't know. Perhaps the ingredients require a fortuitous spark from an outside source to trigger life. But if life can spring from non-life under the right circumstances, then there are likely billions of planets that are in the Goldilocks zone of their sun, have similar chemicals to those found here, and orbit a star with billions of life-sustaining years to go.

Many of these planets are so far away that visiting them is out of the question for the foreseeable future. Even sending a signal to some of these regions would take millions or even billions of years, with a return signal taking just as long. So while the chance of life existing elsewhere in the universe seems very high, the chance of our encountering it is much less.

Finding any form of life within our own solar system would tell us a lot about the value of continuing our search. If life has had a chance to exist elsewhere than on Earth, then it has a chance on many different planets. If life requires something more than chemistry and physics, then we may indeed be alone. It is exciting to be living in a time when we may be getting the first answers to those questions.

E.T.

The search for extraterrestrial life has taken a few different paths. One is called SETI (Search for Extraterrestrial Intelligence), and for over fifty years it has spurred groups to listen for signals from outer space that infer an intelligent source. The SETI Institute was founded in 1984 and formalized the broader search efforts. According to its website, "The mission of the SETI Institute is to explore, understand and explain the origin and nature of life in the universe and the evolution of intelligence."[7]

One of the original promoters of SETI was Frank Drake, an astronomer with the National Radio Astronomy Observatory in Green Bank, West Virginia. He devised a formula in 1961 to help predict the number of other intelligent life forms in the universe that might be a source of communication. Less of a precise calculation and more of an agenda for exploration into those factors that would make signal-sending life elsewhere more probable, Drake's equation includes such variables as:

- The rate of formation of stars suitable for the development of intelligent life.
- The fraction of those stars with planetary systems.

- The number of planets, per solar system, with an environment suitable for life.
- The fraction of suitable planets on which life actually appears.
- The fraction of life-bearing planets on which intelligent life emerges.
- The fraction of civilizations that develop a technology that releases detectable signs of their existence into space.
- The length of time such civilizations release detectable signals into space.

Much progress has been made in assigning numbers to some of those variables, but no other form of life—intelligent or not— has yet to be found, so there is no real measure of how common intelligent life might be.

Searchers have recorded a lot of false positives over time, but no confirmed detection of an intended communication. Despite the amount of time that SETI searches have been operating, they have only been able to explore the tiniest portion of the universe and have only listened at a limited number of radio frequencies. And even 50 years is a very short time to happen to intersect a signal from another world. Signals from far away take eons to reach us. Civilizations on other planets may be intelligent, but not yet capable of sending or receiving radio signals. Or perhaps they have gone beyond that technology, having themselves searched for intelligent life on Earth a million or so years ago and given up after not finding any. SETI is intent on continuing its search. It only takes one confirmed contact to open up a universe of possibilities.

Exoplanets

Another approach to exploring the possibility of life on other planets is to find planets likely to support any life. Planets outside those in our own solar system are relatively small compared to their visible stars and far away. Hence, it is difficult to see them even with our most powerful telescopes. But we can detect their presence in other ways, and the technologies to do so are constantly improving.

Planets have mass and hence exert a gravitational pull on the suns they orbit. The elliptical shape of those orbits mean that the planets vacillate between being closer and farther from their sun. This can cause a slight wobble in their sun's position, and that wobble may be detectable by some of our space-based telescopes. We know a planet is there, even though we can't see it.

Another more common way of detecting a planet is to use a telescope capable of precisely recording the brightness of an observed star. If that brightness lessens ever so slightly, it may mean that a planet is crossing in front of it relative to the telescope's point of view, blocking a tiny amount of the starlight usually emitted. If that dimming occurs on a repetitive basis, it probably means there is a planet orbiting the star. You can get a feel for what the space telescopes are measuring by observing the larger moons of Jupiter through a backyard telescope. The moons cast a shadow on the planet as they pass in front of it, thereby slightly dimming the light you would see if no moon was transiting. A profound example of dimming occurred in 2017 when our moon passed in front of the sun, resulting in a total solar eclipse as viewed from Earth.

The time between repeat-dimmings of a star provides a measurement of how long the planet takes to complete an orbit (its year). Knowing the amount of dimming gives a clue to the size of the planet. Knowing the size of the star can provide further information about the planet, since gravity is based on the mass of the objects and how far they are from each other.

Over 3,500 exoplanets have been confirmed, with many more candidates being considered. However, only a few of these are close to the size of Earth and within the habitable zone of their accompanying star.[8] In addition to its abundant liquid water, the Earth has a number of other features that contribute to and protect our abundance of life, such as the heat generated from the molten core within the Earth and our magnetic poles that divert radiation. It remains to be seen whether those features are essential for all life.

A planet can be kept warm by being close to a small star, or an Earth-distance from a star like our sun (93 million miles), or even further from a larger star. The successful search for exoplanets has mainly occurred within the past few years, so this type of exploration is just beginning.

One of the more promising recent discoveries has been the 2017 detection of seven planets orbiting a star called TRAPPIST-1, about 40 light-years from Earth.[9] The system was named after the Belgium telescope, operating in Chile, that discovered it, but also alludes to a famous beer named after the Trappist religious order. The sun of this solar system is relatively small—a red dwarf—and cooler compared to our sun, but the planets detected are much closer than those in our system and have very short years. It appears to be a healthy sun with a long life expectancy. All of the planets are thought to be rocky like Earth, rather than

TRAPPIST-1 with its seven planets: b,c,d,e,f,g,h, including two passing in front (not to scale). None of the planets are visible in our current telescopes. Planets e, f, and g are in the "Goldilocks zone."

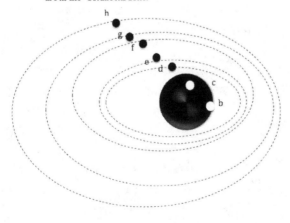

Fig.4

60

gaseous like Jupiter, and at least three are in the sun's habitable zone. It is likely that water exists on at least some of the planets. Some SETI explorers have already started listening for signals from the TRAPPIST-1 system. If intelligent life exists there and if it has developed the technology to receive and interpret our electromagnetic signals, it may be starting to view television shows from the 1970s. In terms of responding to our signals, comments from them about our radio programs of the 1930s might be just reaching us now.

Chapter 5

Time and Space

Thinking of the entire history of the universe all at once, rather than thinking of the universe as a set of things that are constantly moving around, is the first step toward thinking of time as "kind of like space."
– Sean Carroll[1]

Thinking about time can give you a headache. Most of us would divide time up into three realms: the past, the present, and the future. The future is unknowable. We can plan for tomorrow, but there will always be surprises. The past is a little clearer. It happened, it's done, and you can "look it up." But how well do we really know history? Our memory only covers our own life, and is not always reliable. We can reach a little further back by listening to the stories of our elders, as they did with their elders.

Our memory of our past can be both incomplete and subjective. Even a written account is filtered through the individual mind of the writer, so its narration of history is partly subjective. It's helpful to have many perspectives on what was happening at a particular time. Books (or other media) are an important part of our broader history. If no one collects their memories and those of others and records the lessons learned from a time period, that history may be lost forever.

That leaves the present. It is something we experience personally and can talk about reliably. How much time is there in the present? The precise moment that we are experiencing now just became the past. The anticipated moment we are about to encounter is in the future. Because it's so close, it is likely to happen as we imagine, but it's still not knowable. The present has no duration in time. I can feel the headache starting.

The Arrow of Time

Time is the most mysterious dimension. The dimensions of space could care less about humans, but time seems to be intimately entwined with being alive. As far as we know, time is simple, having only one dimension, while space has three. Even in the strange world of string theory—the latest comprehensive attempt to explain our material universe in fundamental terms— there is only one dimension of time, while there may be ten dimensions of space. And time seems to be going in only one direction, while from any point in space you can go forward or back in any dimension. The single direction of time is something we can relate to. As time passes, both the universe and human beings are getting older together.

The laws of science are often embodied in mathematical equations. Using Newton's laws of motion, you can accurately predict how fast a ball will be traveling when it hits the ground if you know how long it has been falling. Many of the equations of science involve plugging in an amount of time. Generally, those equations work fine regardless of whether the time you put in is positive or negative. But we never see time go backward, or home runs popping out of the grandstand and flying back to the hitter's bat. Such events are allowed by our equations, but it's almost as if the equations were meant for a reality other than ours.

Early myths pictured the realms of the gods as timeless and unchanging. Some held that when we died, we went up to heaven where we would live forever. Heaven may be anywhere, but it does not appear to be among the stars, which are constantly changing and time-bound. As a star or a galaxy progresses through its stages, it never goes back, but rather pushes on to new states. Stars are born, stars grow old, stars die. Everything is in flux on a time line, but only in one direction, while in space some planets spin clockwise and others go counterclockwise.

Entropy

There is one branch of science, however, that does follow the unique arrow of time. During the industrial revolution, scientists observed that there is no such thing as a perpetual motion machine. Some energy is lost whenever it is used to perform work. The generalization of this concept is captured by the Second Law of Thermodynamics, which, in a simplified form, says that disorder is always increasing. Of course, many things can temporarily become more ordered, from your sock drawer to the building of the International Space Station. Human beings are a highly ordered arrangement of atoms compared to a random collection of particles.

If we want to make things locally more orderly, we have to do work and use energy, thereby causing greater disorder on a wider scale. Digging and burning coal uses up energy and emits disorderly gases. Without a continuous infusion of energy, whatever we carefully arrange will fall into decay. Because the location of particles in the subatomic world is inherently random, there is a slight chance that a collection of particles will be temporarily found in a more orderly state than normal. It's statistically possible that the cream you stirred into your coffee will temporarily all congregate at the bottom of the cup, but the odds are enormously against it.

Science uses the term entropy to designate the level of disorder and has even devised a way to quantitatively measure it.[2] To say that the entropy of the universe is always increasing is another way of saying that time marches on.

Measuring Time By The Expanding Universe

The universe itself is aging in a way measured by its expansion. This came as somewhat of a shock to scientists who thought the universe was static. It wasn't discovered until 1930 by Edwin Hubble,[3] after whom the Hubble Telescope is named. The distances in outer space are so vast that it is hard to detect movement, even though things are moving very fast. You have to really concentrate to see the moon move as it orbits the Earth. It's actually traveling about 2,300 miles per hour. You know it will be at a different place in a few minutes, but it's hard to detect the movement.

And the moon is very close to us. The stars, especially the stars in other galaxies, are vastly further away. So it took some careful measurements to detect that they were moving away from us. Usually, you can tell something is moving away because it appears to get smaller over time. But these stars are already so far away and the time over which we observe them is so short, that they just appear as points of light with no measurable change in size.

Hubble and other astronomers measured the movement of the stars by noticing that light from certain familiar stars was "red-shifted" compared to a star moving more slowly, like our sun. The chemicals in a star emit various forms (wavelengths) of light, and each chemical has a particular signature—bands of color on a visible spectrum that can be studied. Light of a higher wavelength has a different signature and color than light of a lower wavelength.

A quick review: Light moves predominantly as a wave. To picture a wave, recall the earlier example of a rope that you wave by moving the end up and down (Chapter 2). You can increase the number of high points along the rope by waving the end more rapidly. Such a wave has a higher frequency. (Graph A, compared to Graph B). The distance between one up-portion and the next up-portion will be shorter if the frequency is greater. The wave is more tightly squeezed together. The distance between two up-portions (crests) is called the wavelength. The speed of a wave (that is, the speed at which a crest moves horizontally along the rope) is simply its wavelength multiplied by its frequency (the number of wavelengths per/second; hence, the result is the distance traveled per second). Since the speed of light is constant, the higher the frequency, the shorter the wavelength.

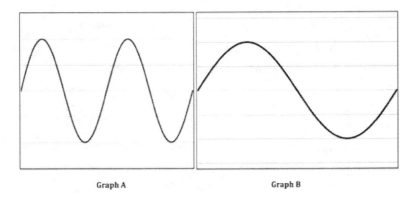

Graph A   Graph B

Fig.5

If a star being observed is standing still, relative to the Earth, its light would have a certain reading on the color spectrum determined by its wavelength. Short wavelengths appear bluish; longer wavelengths tend to the red end of the spectrum's rainbow. Hence, if the light from a star has its wavelength elongated by the fact that it's constantly receding from our view, it will appear redder than it ordinarily would have if it were stationary. Hubble knew what a static star's light should look like through a spectrometer. What he observed in distant galaxies was light that was shifted further toward the red end of the spectrum, meaning that the stars he was observing were flying away, and at a pretty good clip. (Doppler radar is used in a similar manner to catch speeders. Radio waves bounced off a car approaching the radar gun above the speed limit will have a "bluer" shift than cars approaching at a slower speed limit.)

Strictly speaking, Hubble wasn't just observing a fleeting star. All similar stars appeared to be receding in the same way, with the further ones receding more quickly than closer ones. He concluded that it was space itself that was expanding and increasing the distance between him and the stars. We think of space as a vast emptiness, peppered here and there with stars and planets. But space itself is "something." Scientists have not concluded whether or not it is infinite, but our observable universe, the one in which light has had time to reach us, is expanding.

The Limits of Space and Time

Space and time are all around us and within us, but both are hard to define without using the concept of space or time within the definition itself. They are also hard to describe scientifically. Einstein concluded that space and time are intricately related, and that there is no absolute measuring rod to determine distance or universal clock to mark the passage of time.

The Truman Show was a popular movie some years ago starring Jim Carrey. Truman (Carrey) didn't know it, but he was living in a vast Hollywood set. His world looked real, but it was actually just a reality show, with much of the country watching his every televised move. Truman begins to have suspicions that his perfect life was being controlled. He gets into a boat and starts heading out in an attempt to test his freedom and the world outside his town. Since the sea he is on is, in reality, very small, he eventually bumps into a wall made to look like the horizon where the sea and sky appear to come together. He's reached the edge of his space. When a door opens on the wall of the "sky" he begins to understand his situation.

Science doesn't believe that space ends in a wall, even though it may not be infinite. One of the early arguments against space being infinite is that the night sky is mostly dark, beautifully sprinkled with diamond-like stars. If space were filled with an infinite number of stars, between every two stars would be another one, perhaps filling the whole sky with light. It would be like looking up into the glare of floodlights at a nighttime ball game. The fact that the night sky is mostly dark may mean that the number of stars in our universe is finite. In addition, we can only see the stars whose light has had enough time to reach us, and that time is limited by the duration of our universe's existence, some 14 billion years. Stars also die out, with some becoming black holes, which we definitely can't see.

In any case, our observable universe, the one we can reliably talk about, is not infinite. It is expanding, and we can calculate its path back in time and size to a moment about 14 billion years ago when the universe was extremely small. To produce the universe that exists now from that tiny dot required a sudden and powerful burst of space: the Big Bang and its aftermath.

The fact that the light from some stars takes over 13 billion years to reach us is a measure of both the vast size and the age of the universe. Light travels very fast, much faster than sound, which is why we see the lightning before we hear the accompanying thunder. Nothing can travel through space faster than light. It covers a mile in much less than a second. In fact, it traverses 186,300 miles in one second. For most things we see on Earth, that is virtually instantaneous. But if the distances are great, we can notice the time light takes to travel that distance. For example, light from the sun takes about 9 minutes to reach us because it has to travel about 93 million miles, and even for light that's not just around the corner. When the weather report notes the time of sunrise, it is giving the time when the sun becomes visible in the east. It's actually been peeping over the horizon for about 9 minutes—we just couldn't see it (or feel it).

After the sun, the next closest star is so far away that its light takes over four years to reach us. Most stars are much further away than that. And some are so far that it takes a billion years, even traveling at a speed of 670 million miles per hour. The fact that light from the furthest stars took about 13 billion years to reach us gives a lower limit on the age of the universe. If a star existed 13 billion years ago, the universe must have existed at least that long.

As indicated earlier, because the universe is expanding, its radius is larger than the approximately 13 billion light-years that light has traveled since it escaped from the big bang. The detection of the distant galaxy noted above that was formed about 13.4 light-years ago is actually at a "proper distance" of 32 billion light-years.[4]

Before Time

There is no evidence that the universe is headed to an abrupt halt in a few billion or even a trillion years. Instead, it seems the universe will keep expanding, and gravity will become less and less influential. Since the rate of expansion is accelerating, the future may be pretty dim, as the stars recede and become fainter and then invisible to us and to each other. The 14-billion-year history of our observable universe may be just a drop in the bucket of time from that perspective. Our days as a species on Earth are fragilely dependent on the sun, air, and water, all of which will be ended by other changes in the universe. Our sun will use up its fuel in a relatively short time, about 5 billion years from now, dying at the typical (for a star) age of 10 billion years.

What was going on before the universe, and what's going to happen after all the stars stop shining? It used to be that inquisitive minds were told not to ask that first question: what was happening before the big bang? The laws of physics did not exist before the big bang, so there was no way to talk about that subject. It was before time, so there was no time before the big bang. If there was absolute nothingness for infinitely long before the big bang, then time loses its meaning. You can measure time once the universe has popped into existence until it somehow stops, but time is meaningless in eternity. Clearly, our time-trapped minds cannot really wrap around these concepts. The duration of our visible universe's existence is just a blip. Once it started, clocks of various kinds started running, but only to measure the relative events that were taking place: the expansion of the universe, the birth and death of a star, the important milestones in your life.

It's hard to leave a question alone once it's been asked. There are some theories about what came before the big bang, even though our vocabulary is deficient. Perhaps other universes have been popping into existence forever. They emerge out of nothing, they rapidly expand, they contract back into a big crunch, and so it goes. The science becomes very speculative.

Of course, it is doubtful that any human creatures will be around to observe much of this. As previously observed, our days as a species on Earth are fragile.

It would be interesting if time had another dimension, so that if our time expired in one dimension, it still might continue eternally in its other dimension. Some scientific theories allow for more than three dimensions of space. It's difficult to imagine another spatial dimension, but that may be narrow thinking. Basing reality only on what we can see is a surefire way to avoid progress.

So why not four, five, or even ten dimensions of space? If we were one-dimensional, our lives would be boring. We could only move along a line, back and forth. The line need not be straight, but we couldn't go off the line. There could be a big two-dimensional world out there, but we could only see along our narrow line. Very dull.

The same can be said of the limited life in a two-dimensional existence, such as the creatures in *Flatland: A Romance of Many Dimensions*. In two dimensions, our paths could go north, south, east or west, but not off the plane of our universe. Talk of a third dimension would seem absurd. Life exists only on an extended piece of paper. From our actual three-dimensional world, we want to shout at the people in the flatlands to look up, or bore a hole in their surface, but to no avail—everything is two-dimensional—depth has no meaning. Three dimensions are infinitely more interesting. Now things have a solid feel. We can fly, or drill to the center of the Earth. There are stars above us. Life is good in three dimensions and space is now complete.

Of course, it doesn't have to be that way. There could be four-dimensional folks out there clamoring at us to see that added dimension. It would open up whole new worlds to us. Three dimensions are boring. But we can't see beyond our world. If we could get beyond three dimensions, then going to ten doesn't seem such a stretch, especially since the added dimensions are apparently curled up pretty tightly. Like the ant walking back and forth but not around the telephone wire, we might be missing another dimension because it is so small relative to our ordinary world.

Theoretically, we can go wherever we want in the three dimensions of space. We don't seem to have the same mobility in time. We can't go back in time, except perhaps in our memories, and we can't rush into the future. We'll get to some of the future, but only when it comes to us.

The Relativity of Time

Time according to Einstein is just one of the four dimensions of spacetime. He concluded there is nothing absolute about the pace of time. There is no giant ticking clock that determines the number of seconds since the dawn of creation. Rather, everyone's time is relative, and the speed of your clock depends on how fast you are going compared to someone else.[5]

This is a radical notion. In many ways, we depend on time being uniform and independent of the observer. What good is a race if each participant has a different clock, ticking at a different rate? Getting the trains to "run on time" requires that the clocks at the starting point and at the destination are in sync. One of Einstein's brilliant insights into reality is that time is relative.

If you're riding on a train at a constant speed, and you toss a ball straight up in the air, it will come down straight into your hand. To an observer on the ground watching your actions, a different picture is seen. Your toss causes the ball to go up and slightly forward, and you go forward as well, so you meet the ball as it comes down. The ball follows a parabola, the typical path of anything thrown forward under gravity.

There is no mystery here. Your toss provided no speed in the direction of the train, but the train's motion carried your ball forward, as seen by the outside observer. The horizontal speed of the ball is relative to the person doing the observing.

If you stood at the front of the train and threw the ball forward at 10 miles per hour in the direction of the train's travel, the outside observer would measure a faster speed for the ball: 10 mph *plus* the speed of the train. We see this added speed often, as when an athlete takes a running start before throwing a javelin in order to give it extra speed and distance.

However, Einstein realized that light does not behave the same way as a ball (or any other massive object). Light cannot exceed the speed of light, even if tossed from the front of a speeding train. This is a problem. All travelers in space should obey the same rules.

Since the speed of light seemed to be a more fundamental property of the universe than the measurement of distance and time, Einstein proposed that those measurements are not absolute and need to take into account the reference frame of the person doing the measuring. If the reference frame of one observer is moving relative to another observer, their units of distance and time will be lengthened or shortened accordingly.

With a fairly simple set of algebraic transformations,[6] he provided the degree that distances and time are varied if one observer is moving compared to another observer. (An artist also uses transformations when trying to give perspective to a two-dimensional painting of a three-dimensional scene: things in the background need to be drawn increasingly smaller to appear farther away.) At ordinary speeds experienced on Earth, those variations are practically unnoticeable. It's only when one of the observers is approaching the speed of light compared to the other observer that the differences become significant.

Time slows down as your speed increases (and your length is shortened); the closer you get to the speed of light, the slower your watch and your internal aging clock will run, as compared to a stationary observer. Your velocity must be slower than the speed of light, but if you travel really fast, and I stay still, when you return you will have aged less than I. In a sense, you have gone forward in my time.

If you left in a very fast spaceship and rocketed around on a yearlong trip at nearly 186,000 miles per second, your digital watch will say only a year has passed; your onboard computer, which appeared to be working fine, also notes that a year has gone by. But to an observer on Earth it looked like you had the slowest dial-up modem ever. If you left in 2050 and came back in 2051 by your figuring, you might find that the folks on Earth say it is now the year 2100. A lot has changed, but you are only a year older and living in the future.

If you went on a longer journey, you might return to a planet that has advanced thousands of years. Earthlings may have forgotten you ever left. They may even shoot you down as an invader from another planet. Or you may get back to Earth and find that humans have long since gone extinct, or moved to a moon of Jupiter, or been transformed into machines.

This is not just science fiction. This part of Einstein's theory of relativity has been tested numerous times and proved precisely accurate. In one experiment, when a subatomic particle that tends to decay into other particles at a known rate is observed at great speeds close to the speed of light, it decays more slowly than if it were standing still.[7]

Also, using very precise atomic clocks aboard ordinary aircraft going well below the speed of light, scientists have detected slight differences in elapsed time compared to an identical atomic clock on the ground.[8] If we ever do travel at near the speed of light, the differences would be noticeable.

The passage of time is also affected by the relative strength of gravity experienced by two observers. The closer you are to a massive body like the Earth, the slower your clocks will run compared to someone further away. For someone in a rocket ship, this has the opposite effect on time than the effect of traveling fast. Faster speed will mean a slower clock, but being further from Earth will mean a faster clock, compared to those standing still on the Earth. The two effects do not simply cancel each other out, but depend greatly on how fast the rocket is going and how far away it is.

This has very practical implications in our everyday world. Most navigation today, from aviation to a trip in the car, relies on the Global Positioning System (GPS) satellites that are accessible anywhere on Earth. These satellites can send signals to your mobile device allowing it to convey your location and your best route very accurately. They take into account the time for a signal to travel between the satellite and your system. The satellites have onboard atomic clocks that run at a different pace than clocks on Earth because they are traveling faster (slowing down their clocks) and under a lesser pull of gravity (speeding up their clocks). Because gravity is the predominant factor at this distance, the net effect is that the satellite clocks tick faster, and that relativistic difference must be taken into account by the onboard computers so that the location conveyed to you is accurate within a few feet instead of a few miles.[9]

Relativity becomes even more significant because mass (weight) is also affected by speed and gravity. One of the difficulties in achieving faster speeds is that the faster you travel the greater your mass, as measured by those standing still. A greater mass means a greater resistance to achieving higher speeds. Thus, if you want to travel near the speed of light, you will have to have a very powerful rocket—not just because you want to travel faster, but also because you have to overcome your increased weight. As you get closer to the speed of light, your mass approaches infinity. Weightless particles, like photons, have no such problem and can travel at the speed of light—in fact, they are light.

Time Machines

What about going back in time? In the example above, if you and I were both 20 years old at the time of your rocket trip, when you came back I would be 70 years old, and you would only be 21. So, in a sense I would be going back in your time. I could watch you flounder through your 20s, perhaps offer some advice and wisdom from someone who has already been through it. This wouldn't exactly be like going back in time because your 20s would now be experienced in a very different time than I went through. I was in my 20s from 2050 to 2060, you will be in your 20s from 2100 to 2110. So my advice to you will only have limited value—neither of us will have experience in the 22nd century. Your future will be different than it would have been if you had stayed on the ground. You might marry someone close to your own age, not somebody from my generation, even though we used to date in the same circles. But there is no contradiction here—you have one time frame, I have another. Your time frame got slowed down, but you never lived the years you are going through now before, so there is no impact on your future.

What if we could really place ourselves in a prior time, as conjured up in science fiction movies? Serious scientists think it may be theoretically possible, perhaps by traveling through a shortcut in the fabric of spacetime called a wormhole (Chapter 2). But they don't believe that you can interfere with the past in a way that changes the future. For example, you can't go back and kill your parents before you were born. If you could, then you would never be born, and yet you already exist, so that is not possible. It's hard to imagine being present in the past without affecting it, as Marty McFly found in *Back to the Future*. If that's the case, then backward time travel may be impossible. With a powerful telescope, we can, however, see what was happening on other stars and planets millions of years ago since their light (and history) is just reaching us now.

Chapter 6

Why Is There Something Instead of Nothing?

The life of the universe, like each of our lives, may be a mere interlude between two nothings.

– Jim Holt[1]

Looking at the complexity of life on Earth and the wonders of the entire universe, it is fair to question whether this all evolved in a random way or whether it's by design with the guiding hand of a creator. It's impossible for science to answer that question based on our ordinary interaction with the physical world. Even if everything that exists has a logical scientific explanation, there's no guarantee that it's not all part of an intricate plan. Science will proceed with its quest for natural explanations even as believers trust that a higher power ultimately wrote all the laws and created all that is.

Another way of framing the "ultimate question" about existence has been posed as "Why is there something rather than nothing?" This question seems to go deeper than pondering how we arrived at the particular world we live in. It does not take existence as a starting point and then ask whether the wonders of this universe and humankind require a creator. Instead, it renders the difference between a created world and a self-sustaining world as secondary and focuses on why there should be anything at all. This question is no easier than the designer question from a scientific perspective. The whole universe may have sprung from the emptiness of space, but why is there an empty space with so much potential in the first place? Complete nothingness seems like a possible alternative, though obviously not the one chosen.

Jim Holt explores various approaches to this question in his book, *Why Does the World Exist: An Existential Detective Story*. He poses this fundamental question to scientists, philosophers, theologians, and other great thinkers, both contemporary and historical. For example, the cosmologist Andrei Linde contributes the notion that the universe is not as grand as it is often described. He believes that the laws of physics allow that the entire universe could have been sprung from a hundred-thousandth of a gram of matter. Still, you need that flake—it's not nothing. Not surprisingly, Holt does not arrive at a definitive explanation of WHY. In his words, "No one can confidently claim intellectual superiority in the face of the mystery of existence."[2]

What does nothing mean? Scientists will tell you that the vastness of empty space is *not* nothing.[3] Even if you were to choose a place far from any stars or planets, devoid of any particles, that space would still be filled with fields possessing energy and virtual particles that could pop into existence at any time.

Complete nothingness is not to be found in space, it is the absence of everything. Even a black hole has some properties. Nothingness means turning out all the lights and then imagining that there are no lights, no switches, and no one to turn them on or off, and there never was any of those things. Since we humans are clearly something, it's hard for us to imagine nothingness. Nothingness is no longer possible, but it's possible to imagine that nothingness always prevailed—that means no us, no particles, no physical laws, and no supreme being to create it all in the first place. No beings at all, no virtual particles or invisible fields, no time or anything to make a big bang. Just nothing, and nothing to take notice.

Any slight break in that perfect nothingness would be something. Our whole universe could have evolved from a thimble full of energy. Just a bit of empty space with the potential for a virtual particle to fleetingly pop into existence is something. It seems there are endless possibilities for different kinds of somethings, but only one kind of nothing. Nothingness is unique, complete, perfect. Something can have many forms. It can be messy, disorderly, driven, complex, or very simple.

I can think of no compelling reason why there should be something rather than nothing, but it could just be the way things are. But it's also hard to see why nothing would have been the more likely state, requiring unnatural intervention to overcome. If there are a million ways to have something, but only one way to have nothing, then it might be more probable that we find ourselves in a state of something. That alternative seems even more demanding once we observe that something clearly is possible because our universe, even if it is somehow more of an idea than reality, is a prime example of something. Nothingness excludes even the possibility of something. The reverse is not true: we can imagine nothingness, even while we know there is something. Something is clearly a potential state, so perhaps it was inevitable.

The leap from the dreary possibility of nothing to something is a huge one. It opens all kinds of doors. One version of something is a self-propelling universe in an endless state of expansion and contraction, with life thriving at particularly opportune times. Another version is a universe created and governed by an all-powerful God, who has a particular concern for human beings. In a way, that is two somethings: a supernatural being and that being's creation. It is the first something that is most crucial: Why should there be a God rather than nothing? Of course, that question is also unanswerable, but it highlights the key dichotomy: while nothing was a possibility, somethingness prevailed. No particular something—for example, a self-perpetuating universe or God who created and designed the universe—would seem to have preference over another. They are both wondrous compared to nothing. We may never know the answer to these mysteries.

SIMS

There is an odd theory about human existence that renders us more than nothing, but considerably less than the something we like to think of ourselves as—that is, the pinnacle of creation and evolution in the entire universe. With the advent of computers came computer games. A whole generation has grown up immersing itself in video worlds of danger and competition, while accidentally learning skills that might be useful in the years ahead.

One of the goals of video-game makers is to create visuals and action that is as close to real as possible. Because of the limitations of computing power and a full understanding of how humans behave, such games are at best simulations of the real world. Perhaps it was inevitable that futurists began to think of the possibilities that more computing power and understanding of human nature could bring. If we could approximate human consciousness in a computer, then the avatars of our video games would act much like real humans.

From there it's only a short leap to the notion that maybe we are already in that future. Or to be more precise, there are already advanced humans (or another intelligent species) who can create games of which we are merely the simulated characters, acting out our present for the enjoyment of the gamers. Instead of Truman (see Chapter 5) being a human pawn of his corporate and Hollywood manipulators, he could be just a seemingly self-conscious game piece in an elaborate electronic simulation of life in the 21st century. The assumption behind such a game is that people, who are now in the future relative to the SIMS (computer simulations) characters such as us, will enjoy creating super games depicting what life was like in our current times. I'm not sure our lives are that interesting, but who knows?[4]

If computing power is sufficient, we (the characters in the game) might have a degree of consciousness, of free will, of emotion, and everything else that makes one human. Our consciousness would be false, as would our memories. We would believe that we are real, that if we kick a rock it will hurt our toe, and that we come from a long line of other real beings. We would not be mere programmed automatons. The unpredictability of the actors is part of what makes SIMS interesting. Think of the fascination that everyone had watching Truman. Millions of people were glued to their TV sets to see what he would do next. Of course, that was all fiction.

Although this seems like a preposterous idea, people (or are they, too, just SIMS?) who believe in this theory today also believe that we are far more likely to be characters in the game than potential creators and players. The theory imagines that millions of SIMS could be created by just a few developers, implying that the odds of being a game piece were much higher than being a gamer. Also, the SIMS could be so advanced as to create their own sub-SIMS, and so on down the line. We might be a fifth generation game, created by the very intelligent fourth generation SIMS, who in turn were created by the super-intelligent third generation. At the end of the string, there may be just one master gamer pulling all the strings. If you think you're a conscious being, chances are you are a SIM, just because there are many more characters than masters.[5]

Trying to think your way out of this maze is hopeless. Like the existence of the divine, it may or may not be true, but logic or experiment will probably not get you to a certain conclusion. Perhaps it seems unlikely that future humans will care that much about a game reenacting the days of their ancestors (us), but on the other hand, if it can be done it's hard to prove that it won't be done, at least by a few. It's also possible that the game we are in has nothing to do with real society either past or present, and our whole SIMS world is a fanciful creation of an alien species.

We are perfectly free to reject this whole notion and continue to live our lives confident we are independent, fully conscious human beings with actual free will and emotions, living in a physical world on a real planet called Earth. But of course, that's just what the game makers expected us to do.

One of the downsides of living in a SIMS world is that our pursuit of scientific knowledge becomes superfluous. Nick Bostrom of Oxford has written a rigorous piece on SIMS theory, complete with equations and probabilities. He acknowledged the bind:

> If we are living in a simulation, then the cosmos that we are observing is just a tiny piece of the totality of physical existence. The physics in the universe where the computer is situated that is running the simulation may or may not resemble the physics of the world that we observe. While the world we see is in some sense "real", it is not located at the fundamental level of reality.[6]

Elon Musk, the rocket builder and futurist, is also one who believes the SIMS theory might represent our reality, but he sees a bright side for intelligent life:

> We should hope that's true because otherwise if civilization stops advancing, that could be due to some calamitous event that erases civilization, so maybe we should be hopeful this is a simulation.

In other words, the alternative to a civilization that creates SIMS may be no civilization at all:

> [W]e will create simulations that are indistinguishable from reality or civilization will cease to exist. Those are the two options.[7]

Part II

From Telescope to Microscope

The second part of this book will examine the universe on its most fundamental levels. The world under a microscope is not a different one from the one we see at the farthest reaches of space. Stars are made from atoms, and everything we see is illuminated by photons just like the ones bursting forth at the big bang. One set of physical laws should govern both the realm of the very large and the very small, but this grand theory has yet to be devised.

The first chapter of this section looks introspectively at ourselves and what makes life on Earth so successful. The understanding of DNA and its role in all life on this planet is a very recent discovery, compared to the exploration of chemical elements that make up DNA, or even compared to an understanding of the elementary particles that make up the chemicals. Rather than follow an historical trail of scientific discovery, Part II will go from the small, to the very small, and then to the smallest parts of the universe.

Chapter 7

A Code for Being Human

If there is a human moral to be drawn, it is that we must teach our children altruism, for we cannot expect it to be part of their biological nature.

– Richard Dawkins[1]

There is a nature video showing the plight of a large herd of wildebeests in Africa. Every year, their migration to find sufficient grasslands requires crossing a fast-moving river. Being Africa, crocodiles have learned that the wildebeests will be making this dangerous journey and are waiting for them. On the land, crocodiles would have a hard time catching a wildebeest, but in the water they are master predators.

It's tragic to watch what happens. The wildebeests are driven by a strong instinct to find new food and that means jumping into a raging river where the predators are lurking. Fearing their possible fate, they nevertheless take the plunge one after another. The youngest are the most vulnerable, but even mature cows can be taken under water by the huge crocs. In high definition, you can clearly see the terror in the cattle's eyes as they try to reach the other side alive.

Most do make it, since there are far more of them than crocodiles. As humans, however, we are thinking, NO, don't go in the water! You are defenseless and many of you are going to die, including some of your children. Surely, there must be some alternative. What would we do? First, shoot the crocodiles. Second, build a bridge over the river and use it to get to the grasslands. Even a ferry would be an improvement. At a minimum, perhaps conduct a thorough survey of the area to reveal a safer place for crossing.

In all likelihood, the wildebeests have explored the possibility of another crossing. In a herd, there may be one outsider that accidentally wanders down to a narrower crossing point where it is able to rush across without being seen by a croc. Perhaps the beast will remember that crossing point and use it again next year with her calf in tow. The family will grow, and the whole herd might start following the successful cow—but the crocodiles will also adapt.

Among other things, this is a story about genetics. Over the years, the wildebeests may have learned to pick the safest crossing point, even though it is still fraught with danger. Deeper in the wildebeest's nature is the instinct to seek new grass at certain times of the year. It has been ingrained in them through generations as a way of staying alive and multiplying. Collectively, the herd has made a bargain that survival is more likely for more members if they cross the river and get to the grasslands than if they stayed in a dried-out land, or expended too much energy searching many miles away for a safer path. Some will be lost, but most will live and the herd can grow. Survival is the most powerful instinct.

Observing such situations, we pat ourselves on the back with our superior intelligence. We would not just follow the herd blindly into almost certain and horrible death for members of our extended family year after year. We would use tools, creativity, and cooperation to find another way. Building bridges, ferries, or using weapons are not part of our instinct. We have the advantage of having a relatively large, self-conscious brain passed on from one generation to the next, and that is our nature. We can study problems and come up with solutions, rather than accidentally finding a better way.

Lest we become too smug, however, it is important to admit that we often do follow the herd. Newcomers, "different" looking people with new ideas, are often shunned, or worse. New ways that challenge the status quo are not readily embraced. We, too, allow many thousands of our extended family to be killed each year, even though we know very concrete (though perhaps unpopular) ways to save those lives. A higher-intelligence species might disdainfully observe us at times in the same way we observe the wildebeests.

Inheriting Traits

How do animals pass on the skills necessary for their survival? Baby birds never see their nest being built, but they will know how to build one when their time comes. And how do our human qualities get passed on to our children? Although many cultures have beliefs about how animals, including humans, were introduced on Earth, the widely accepted scientific theory is that of evolution. Charles Darwin, a British scientist born in 1809, had some ideas on the emergence of new species, and he gained further insights while traveling the world aboard the HMS Beagle for five years.[2] One of his ports was the islands of Galapagos, off the coast of Ecuador. Since the islands were sufficiently far from the mainland, unique animal and plant species had a chance to grow without interference from more dominant species on the continent.

For example, Darwin noticed that the beaks of certain species of finches differed depending on the food available in various places. He concluded that physical change in animal species occurred gradually, with each successive generation retaining traits that proved favorable. Rather than being the result of simultaneous creation, the finches with different beaks likely emerged from a single species of finch, as a response to the food that was available in a certain locale. Some beaks are good at cracking nuts, others work well on fruits. If a finch finds itself in an environment surrounded by fruit, it is the latter beak that will emerge through successive generations. At first, these new traits appear as anomalies—accidental variations in the usual body type. But if an anomaly proves useful in securing the survival of the bird, that bird will live longer and pass on that fortuitous trait.

The theory of evolution includes the corollary that a whole new species can evolve as the favorable variations within a species make the new version so different as to constitute a new kind of creature—one that no longer reproduces by mating with creatures that did not adapt. Thus, single-celled animals can become multi-celled, and multi-celled animals can develop eyes, brains, fins and lungs. Fish can become land animals, perhaps by first adapting to shallow marshes, and mammals can evolve into humans, if larger brains convey an advantage.

Darwin did not have a sense of how long all of this change might take. Science now believes the Earth was formed from debris circling the sun about 4.5 billion years ago. Simple bacteria emerged relatively soon thereafter, about 3.8 billion years ago. Life remained in a relatively primitive form until about two billion years ago, when single-celled creatures with a nucleus started to form and life began to diversify.[3]

Mammals may have developed in the age of the dinosaurs, about 100 million years ago. Due mainly to an outside intervention (a large asteroid colliding with the Earth in what is now Mexico), the extinction of the dinosaurs took place 65 million years ago. It appears that a few kinds of dinosaurs had the right traits to survive this catastrophe. Smaller dinosaurs that could fly and had feathers evolved into the thousands of species of birds in our world today. If you've ever seen a wild turkey running, it becomes quite believable.

Small mammals also made it through the dark times that encircled the Earth after the collision. Without the dinosaurs, they flourished and expanded their territory. Human-like species (Neanderthals, for example) began to appear about 300,000 years ago. Modern homo sapiens (us) have a relatively short history so far, appearing about 50,000 years ago. We either mingled with or eliminated the other human-like species.

If the entire 4.5 billion-year history of the Earth was proportioned over a single year, then microbes would have appeared around February, sea stars around July, but our species would not have appeared on the scene until about 30 minutes before midnight on December 31.[4]

Two observations can be drawn from our short existence on this planet. One, it took a relatively long time for intelligent life to emerge here. This may say something about the possibility of finding similar life elsewhere in the universe. Every moment of the Earth's existence and evolution's slow trajectory was needed to produce a species capable of communicating beyond our planet. Random mutations had to play out in a very favorable way.[5]

After four billion years of trial and error, we have briefly taken our place on the stage of life. Many things about the Earth's place in our solar system and about our solar system's position in our galaxy had to be just right for this long period of germination to endure. Some massive stars die out and explode after just a few million years of productivity. Our sun has already endured for five billion years. The evolution leading to intelligent life on Earth occupies a considerable portion of the history of the entire universe. There are lots and lots of stars, but a much smaller number with the endurance and the favorable conditions of planets steadily aligned in just the right way for over four billion years, almost one-third of the 14 billion years since the big bang.

The other observation is that a long future for us is in no way ensured. Species come and go. True longevity among some animal species can be measured in hundreds of millions of years; we have been around only for tens of thousands. Our genes are keyed to survival, not necessarily to achieving the highest intelligence. Self-destruction, extinction events, or our evolution into a totally different life form are all possible. It certainly seems likely that, if we survive, we will change more radically in the next 1,000 years than we have in the past 5,000. It took about

5,000 years from the introduction of the wheel to the emergence of the automobile. It only took about 75 years between the first car and astronauts driving on the moon.

DNA

What Darwin also did not know about evolution was the biological mechanism that allowed for adaptive changes in species. Where do our traits reside within our bodies and how are they passed on? How do changes in a species occur, whether it is a larger beak in a species of finch or a larger brain in the human body? The secret for passing on all of the necessary traits for survival to one's offspring—along with a few accidental changes as well—is found in the DNA molecules that reside in almost every cell in every living thing.

DNA is a complex combination of chemicals and stands for deoxyribonucleic acid. The discovery and understanding of DNA is one of the most important developments of the 20th century. Just as computers have given us an invaluable tool to create and manipulate information, the understanding of DNA has given biology and medical science the tools to probe and impact life at its most fundamental level. In fact, the linking of DNA and computers goes deeper than the fact that both represent incredible advancements in human knowledge. It turns out that DNA acts very much like a computer code, providing instructions to the body in long strings of chemical signals that act as a simple language.

We will explore later (Chapter 9) how everything in the digital world can be represented by a series of 1s and 0s. They can convey every number, every letter and word, and even provide instructions for manipulating numbers in complex formulas. The world of biological information works through a similar process. Instructions in the cell are conveyed through messages encoded in sequences of four chemicals: adenine (A), guanine (G), cytosine (C), and thymine (T). When a cell senses the presence of specific sequences of these chemicals, it springs into biological action, putting together the building blocks that the body needs.[6]

When a computer code initiates a new program, it awakens the processor with a special series of 0s and 1s indicating that a command is coming. When we click an app on our smart phone, we send a similar notice, which is translated into the zeros and ones (really switches being open or closed). We then feed in more information, and the app "knows" what to do when it "senses" electric current coming through the appropriate switches. When we quit the app, the computer within the phone receives the command to stop.

DNA works in a similar way. When conditions are right within a living cell, a chemical message is activated in the cell, and it begins "reading" a series of chemicals on the DNA strand. In an oversimplification, that series is what is called a gene. Genes are strings of chemicals that convey instructions using just the four acids that are the code of all DNA. Some strings of chemicals appear to convey no message to the cells. But other strings signal the cell to begin operating, to produce building blocks called proteins, and to help those proteins carry out a special function needed by the body. Those strings are the genes.

The long strand of DNA in a cell (the genome) is made up of a series of chemicals from among the four acids, which can be represented by the letters G, A, T, C. If you were to read the human DNA in a cell using these letters to stand for the chemicals, it would be a string billions of letters long, using only the alphabet G, A, T, and C. But not every snippet of letters along that DNA string can produce meaningful results. Just as a computer may see a big difference between 01010101 and 10101010, there may be a very different message conveyed between GAGAGA and TATATA.

A message to the cell may begin with a chemical string, say G-A-T, that initiates the start of a building process. From that point on until the command chemicals signal the end, the sequence of letters is very important and represents a gene. From the billions of lines of DNA code in each human cell, there are only about 20,000 key messages or genes.

Many less complex organisms like species of worms and plants have longer DNA strands than humans. Scientists recently decoded the genome of the salamander axolotl and found over 30 billion lines of code, about ten times the number in humans.[7] There is more to each species than just its DNA code. DNA is a form of instinct, passed on from parents to offspring. Some species operate almost exclusively on instinct. They need to have a record in their DNA instructions of how every part of their body is made, how to hunt for food, what foods to eat, when to migrate, etc. Axolotls, for example, have instructions in their genes allowing them to regrow limbs and organs.

Humans can leave a lot of that up to education, creativity, and intelligence. We may have an instinctual feeling of hunger that causes us to want to eat. But what we eat is drawn from a very complex array of choices depending on local culture, upbringing, farming practices, and many other variables. We are constantly adapting what we eat and drink in response to medical research (as well as to a barrage of advertising).

So we don't need as many genes telling us things we can figure out for ourselves. As for body structure, we're not that different from many animals in terms of the number of limbs, eyesight, digestive system, brain, and heart. Our bodies have adapted to the environments we choose to live in, but other animals have also adapted in different ways to promote their own species.

Further research into the structure of DNA has shown that each cell in our body does not contain just one long string of DNA arranged in 20,000 genetic messages. Instead, the string is broken into smaller groups of genes called chromosomes. There are 46 such groups in each cell, half of them from one parent and half from the other. Each parent's contribution is similar in that each contains genes for all aspects of the human body, and the similar genes from each parent are located at the same location along their respective DNA strand. The greatest difference among the pairs appears in the chromosome that determines the sex of an individual. Together, biologists say the cell contains 23 pairs of chromosomes.

The basic structure of the DNA within each chromosome is not a single string using the above-mentioned four chemicals, but rather a double string, bound together with chemical bonds. The double string twists in a helix formation, like a spiral staircase. The steps of the staircase are made of "base pairs" drawing from the chemicals G, A, T, and C. One step of the stairway might consist of G and C. Another step might have an A and T pair. In fact, that is the entire rulebook for DNA pairs: Gs can only be paired with Cs and As can only be paired with Ts.

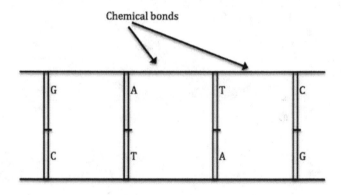

String of DNA four base-pairs long

Fig.6

To picture the shape of the DNA molecule, you might think of a curly ribbon, stretched out by holding both ends. It twists around like a corkscrew. The GATC bonds might also be pictured as a zipper, with complementary sides that can be locked together or unzipped into separate strands.

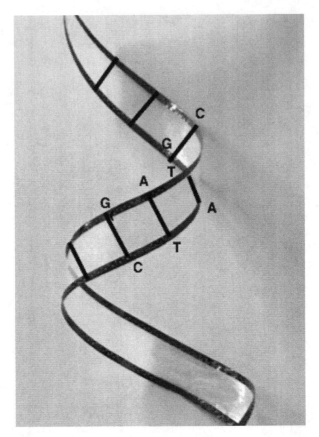

Fig.7

Returning to the spiral staircase image, as the steps go up, they are held in spiral formation by a substructure comparable to the supports on the sides of the staircase. In DNA, this structure is made up of other chemicals that are not part of the fundamental information code. Let's call the two sides of the staircase "left" and "right," even though the original DNA is a twisting helix. If you were to read the sequence of letters on the "left" side of the steps, it would sound very different from reading the right side. For example, if the first four letters on the left side read G-A-T-C, then the first four on the right side would have to read C-T-A-G. (See illustrations.) So if you know the sequence of chemicals (or letters) on one side of the step, you can completely determine the order of the letters on the other side. In that sense, the strands are functionally identical. If there is a string of three Gs on one strand, there will be a string of three Cs on the other strand. This becomes very important when a cell divides and needs to have identical strings of DNA in its daughter cells.

Making New Cells

Before a cell divides, it needs to make copies of its DNA, all 23 pairs of chromosomes. Every cell, whether it is in our toe or our heart, contains the complete DNA instructional manual for the entire body. But in a string of chemicals billions of steps long, and where the order of the chemicals is critically important, a lot of mistakes could happen in an attempt to simply duplicate the chain as a whole. DNA accomplishes this critical task by having the original DNA double helix (broken into chromosomes) unzip into its two component strands.

Each single strand has to now find the proper chemicals to match up and form two double strands. They do this by drawing the appropriate chemical matches from a pool of the four chemicals (GATC) in the cell: every G needs a C-match, every A needs a T-match, etc. The "left" strand of the zipper will gather together the chemicals for a matching "right" strand, and the "right" strand of the zipper will similarly assemble the chemicals for a proper "left" strand. Each single strand contains the exact sequence of chemicals necessary for all the instructions the body needs. The result of the re-pairing will be two double helices, each exactly like the one that had unzipped (though occasionally isolated mistakes still occur). Each of the pairs of chromosomes in the cell accomplishes this doubling, producing a total of four DNA sets, and the cell can then split into two new cells, each having two DNA sets, just like the original cell before division. Each new cell will have identical instructional manuals contained in their nuclei.

There is one important exception to the cell division process described above. Ordinary cell division (mitosis) is basically a form of cloning, but when a reproductive cell (sperm or egg) first divides (meiosis), it only has one set of the 23 pairs of chromosomes. A second pair is added through fertilization of the egg by the sperm, each of which had only one set of chromosomes. The resultant fertilized cell again has a full complement of 23 pairs or 46 chromosomes, half from each parent, just like other cells. Finally, a new genome is then created, using snippets from the sperm (father) and snippets from the egg (mother). The new DNA in that cell becomes the blueprint for the child and all of its subsequent cells.[8]

This contributes to the uniqueness of each individual. Even children from the same parents have different DNA because the various snippets of DNA from the mother and father are randomly recombined in the fertilized egg and thus are different in each child. If a trait like Huntington's disease is determined by a single gene, the child may receive it from either its mother or its father.[9] (Some traits require that both the father and the mother have the gene for that trait.) If a trait like skin color is determined by a number of genes, the child may reflect a mixture of the characteristics of each parent.

The actual working of DNA within the cell is amazingly complex and still only partially understood. The summary above barely scratches the surface of the intricate ballet that has evolved over billions of years to sustain life. The important point moving forward is that science has a deep understanding at the molecular level of how our bodies (and every other living creature) are made and how our species evolves through reproduction. This knowledge opens the door to finding connections between certain diseases or physical defects and corresponding genes in our DNA. It is possible to predict early in life whether an individual will have a defect by examining the DNA in any cell and it may be possible to fix that DNA and extend life. Moreover, since the DNA code is a sequence of chemical molecules, scientists can produce those same molecules artificially and pair them in the same chains, imitating the basic functions of life. This power raises critical ethical questions for the future.

DNA science has proved very useful in other ways. It has been immensely helpful in tracing the evolution and extinction of various species and in allowing people to trace their biological ancestry. It is also playing an increasing role in the criminal justice system by helping to convict the guilty and exonerate the innocent.[10] Because each person's complete genome consists of billions of lines of code that can be arranged in almost countless ways, no two people are likely to have the same DNA (even identical twins, who are essentially clones of one another, are likely to have different mutations). Whether or not an accused was at the crime scene can often be determined a lot more accurately through DNA testing than through eyewitness identification. As we search for other life in the universe, it will be interesting to see whether the DNA model of transferring traits is found in alien species, or whether another model also works.

Survivalist Genes

Richard Dawkins published a fascinating book in 1976 called *The Selfish Gene*.[11] The title could be taken a number of ways, but he was not claiming the discovery of the one gene in the human genome that makes us selfish. Rather this is a book about the evolution and "goal" of every gene in existence. Genes are the formulas for all the parts and traits of every living thing. The selfish gene is not some evolutionary mistake that allowed greediness to creep into our DNA. Selfish traits are the only ones that can survive in the long-term. In a competitive environment, they are the ones that give us, or any other species, a chance to obtain the nourishment to sustain life and the means to reproduce.

Mutations, Good and Bad

After billions of years of evolution, almost all genes are likely the result of some mutation that occurred in an earlier version of the gene. DNA, and hence the genes it contains, can mutate in many ways. Some accidents occur when a cell divides and produces two clones of itself. The double helix DNA splits into two strands, and each strand binds up with the appropriate chemicals in the cell to make two new helices. But mistakes can happen in assembling the mirror string, billions of letters long. Sometimes a G gets matched with an A instead of a C. There is a cellular mechanism for "spell checking," but it doesn't catch every mistake.

Mutations can also happen in the reproduction of living things. In many species, male and female cells combine their DNA and multiply into offspring. Errors can creep into the process. Changes in DNA can also occur because of outside influences, such as radiation from the sun or cosmic rays from outer space. A mistake means that the next generation of cells, and possibly the next generation of the species, will have a different trait than their ancestor. New traits can also emerge because combinations of the male and female genes might allow a trait to appear in a child even though it was not operative in either parent. That trait could be detrimental, such as the trait associated with sickle-cell anemia or with hemophilia. Sometimes a detrimental gene can survive because it also conveys some positive benefits. The gene that determines sickle-cell anemia apparently arose in Africa as a mutation that helped prevent malaria.[12]

Beneficial changes can fortuitously give an offspring an advantage to deal with the ever-changing environment into which it is born. If the trait makes it faster or smarter, it might help the offspring live longer and produce many more generations, some of which will inherit that positive trait. That's the "selfish gene" at work.

We may think of selfishness as a moral failure, but in biology, it is just a matter of probability. The gene possessing a mutation that helps it take better care of itself by finding more nourishment will live a little longer and reproduce a little more. It will survive longer than similar genes without the mutation. Of course, the gene itself has no independent life. It resides on the long string of DNA and is called into action by the cell, which gets requests from the rest of the body. But if that gene, either in its original or slightly mutated form, helps the body to survive, that gene will be carried along with the successful reproduction of its host. The aberrant gene will become predominant, though certainly not immortal, as the process of evolution has no end, except the exhaustion of the resources for life.

We like to think of the human species as somehow escaping this law of the jungle. We know we can be selfish, but we can also be self-sacrificing, generous, and committed to the greater good. We have free will. Our self-consciousness may bestow on us qualities that are not bound to our biology. Our spiritual side, if it is something in addition to our physical makeup, may be capable of directing actions and emotions that transcend the simple daily grind to eat and multiply.

Altruistic Genes?

Dawkins, however, makes the case that the operation of the selfish gene can even be seen in the noblest actions. For example, a mother sacrificing her life for her child is a way of ensuring the passing on of her genes to posterity. Yes, she dies, but that was going to happen anyhow. The important thing is that the child (possessing half of the mother's genes) lives on.

Of course, a soldier might give her life for a stranger's child or for the good of the country. There appears nothing selfish in that. Dawkins might say that protection of the community or even of the whole country may also contribute to the survival of the soldier's genes. In other words, those who have the propensity to throw themselves into the defense of their community have a better chance of extending their genes than those who don't. The soldier may die as a patriot, perhaps even without having surviving children. But her siblings or parents might live longer because the community has been saved, and a portion of her genes reside in those siblings and parents. Those genes have a chance of carrying on.

Genes do not choose their paths or focus on a single action. They are a set of instructions, tweaked by trial and error, where success means moving on and failure means being replaced. Our selfishness and selflessness are part of what makes humans so successful. The genes of our species get passed on, and with that is also the probability that each individual's genes get passed on—and that is what every gene "wants" to happen.

The Selfish Gene is not about human genes; it's a theory about genes in all species. In many ways, humans are not like other species. We have an enormous capacity for empathy, love, and even hate. We enjoy art, music, and sports. We form clubs, charitable organizations, governments, even a United Nations, with groups of people who are unrelated, except in the most remote ways. Our organizations and governments have rules designed to promote the greater good, not necessarily the perpetuation of anyone's genes, and they are not restricted by the patterns of evolution.

Of course, it's possible those human characteristics are precisely what give our selfish genes the best chance of survival. From the genes' perspective, living creatures are merely the "survival machines" for the perpetuation of the gene. Good survival machines with a high chance of reproducing are favored by the gene-driven evolution. So much of what makes us human also makes us good survivors.

Humans are superb toolmakers. Some birds pull grubs out with sticks, and apes smash nuts with rocks, but humans have taken machines to a whole different level. Our machines can beat us at chess and soon will be driving our cars better than we ever did. It is already becoming hard to distinguish machines from life forms. When we look for life, we look for things that can nourish themselves and can reproduce. There's no reason machines can't be programmed to do the same.

A machine's search for food might require no more than seeking light and converting it to energy. Broken parts could be replaced by new parts if the machine is programmed to know how to manufacture the necessary replacements. Reproduction can also be broken into mechanical steps.

Humans have constantly benefitted from evolution, which is a way of interjecting "experiments" into the replication of our genes. Geniuses are not the result of cloning. Machines, too, can be taught to learn from mistakes and to try new models, so that each generation of machines is an improvement on the one before.

Instead of the selfish gene, the future might belong to the selfish machine.[13] Once set in motion with the right technology and the right set of startup instructions, the machine world should be able to go on quite well, independent of "living" creatures. At this point, we have the power to limit what our machines can do, so that they serve our needs but can't decide to replace us. But the best machines will be the selfish machines, the ones that can take care of themselves and ensure their own future through constant trial, error, and the survival of the fittest. DNA may become obsolete—perhaps it already has on other planets.

Chapter 8

Our Common Ground with the Universe

Ultimately, the very tiny objects we study are integral to discovering
who we are and where we came from.
– Lisa Randall[1]

We turn now to structures smaller than the molecules that make up our DNA. Molecules are made from chemicals and ultimately from atoms, which reside at the visibility limits of even our most powerful microscopes. Atoms, in turn, are made of even smaller parts, where our common notion of particles breaks down. Some fundamental particles have no mass, weighing as much as a thought or an emotion. Both life and other matter share the same set of chemicals and elementary particles that apparently exist throughout the universe.

Our eyes evolved to see things like food and predators in order to survive. Seeing distant galaxies or the atoms that comprise our physical world was not essential to our existence. Once we gained greater control over our environment, humans had the time and a burning curiosity to explore the world around us and the world inside us. We have developed powerful tools to help us see both the stars outside of our view and the very building blocks of our universe. And when we reach the limits of our instruments, we turn to indirect ways of measuring the world by the effects they have on things we can observe, both large and small.

Atoms

The ancient Greeks tendered an insightful guess about the structure of material things. They theorized that as you split an object into smaller and smaller pieces, you would reach a physical limit. You cannot divide a substance indefinitely and still have a piece of that substance. Rather, at some point, you would arrive at an indivisible particle that still retains the essence of the object being divided. They called that elementary building block an atom.

The Greeks did not have the technological tools to take this notion beyond a theory. The Greeks had many theories about the physical world, some of which were on the right track and others that turned out to be wrong. Without a way of reliably testing their conjectures, their pursuit of truth was more philosophy than science.

The scientific method of exploring the world blossomed in the Renaissance as we invented tools like telescopes, microscopes, and established standards of measurement. Now theories could be put to the test. For example, the widely accepted theory that the Earth was the center of the universe didn't hold up under the observations made through telescopes and the calculations of astronomers.

The Greeks' atomic theory was also put to the test. The outcome was unexpected, and we have still not plumbed the mysteries of the world at its most elementary level. There are such things as atoms of gold or lead that cannot be broken down further and still retain the properties of gold or lead. But they can be broken down into even smaller particles called electrons, protons, and neutrons, and these particles make up everything we can presently see in the universe. Protons and neutrons can even be broken down more finely, but pretty soon we must abandon our whole notion of smaller and smaller particles. The world of the smallest parts of the universe is very strange.

Identical Blocks Make Different Substances

Alchemists of the Middle Ages spent countless hours trying to transform one chemical element into another, hoping perhaps to make gold from lead or other substances. They had little success. It seemed that, at the fundamental level, gold was made from very tiny gold particles and lead was made from tiny lead particles. It turns out, however, that the alchemists were not totally off base.

One of the many things we tend to take for granted is the astounding interconnectedness of all the chemicals in our universe. Almost all of the things we see around us are made up of combinations of chemical elements. Water, for example, is made up of the elements hydrogen and oxygen. Gold is an element itself, not a combination of other elements. If we look at gold and oxygen, we might say they have nothing in common. Yet they are both made from the same three building blocks. The main difference between these two elements is simply the number of each of the three building blocks used.

The periodic table of chemical elements seen on the walls of high school chemistry classes is an effort to group the various chemicals according to their physical properties. The table contains a series of columns of chemicals under such categories as inert gases and reactive metals. The rows of the table represent different weights of the various elements, with weights increasing as you go further down in the columns of the table.

As scientists discovered new elements and added them to the table, it was noted that some chemicals had a curious behavior. They appeared to emit radiation (heat and light) that could expose photographic plates. As these elements decayed, they gradually evolved into other chemicals. This was one indication that perhaps all chemical elements could be broken down further.

Scientists also found that if they fired a fine stream of atomic particles at a piece of gold foil, almost all of the particles passed straight through the foil. However, a few particles appeared to ricochet off something very small but "hard" within the foil. It seemed that even though gold was a solid metal, its structure was mostly empty space, peppered with tiny nuggets of solidity.

These islands were designated as the nuclei of the gold atom, one of the key building blocks of every chemical.[2]

Gradually, the remaining structure of chemical atoms revealed itself. The nucleus of every element was made up of discrete particles, and on the whole had a positive electrical charge. The simplest element is hydrogen, which has just one particle in its nucleus, a positively charged particle called a proton. Further measurements of the weight of various elements revealed that something else was contributing to the weight of the nucleus besides the positively charged protons. That something were the neutrons—particles similar to protons, but with no charge. Neutrons added weight to the atom, but did not upset the balance of charges.

Around the nucleus, tiny electrons with a negative charge orbited at a great distance, compared to the size of the particles involved. Hydrogen had one electron, but other elements had multiple electrons, and for a particular atom, the electrons were not all at the same distance from the nucleus. That would prove very important in understanding the world of the very small.[3]

The protons in the nucleus attracted the oppositely charged electrons in much the same way that the planets are attracted by the sun (although the attractive forces are very different). The neutrons had no charge. This picture might have seemed familiar to early scientists. Each chemical element appeared to be a microcosm of the solar system. There was harmony in such a structure, with each atom having an equal number of positive and negative particles, and elements differentiated by the total number of their particles.

This neat picture enabled chemists to fill out the periodic table. The atomic structure of new chemicals could be predicted by adding a proton and perhaps some neutrons to an existing chemical. Additional protons were always balanced with an equal number of electrons.

The amazing thing about this structure was that it was built with identical fundamental particles. Each proton is the same as every other proton, and the same is true for electrons and neutrons. Yet when you varied the number of these particles, totally different substances emerged. Neon gas is very different from Sodium—a basic element of table salt—but their nuclei only vary by one proton (and a few neutrons). Every chemical on Earth, and perhaps in our known universe, is made up of the same three particles, just in different numbers.

For the alchemists, gold is structurally not that different from lead. They differ by only a few protons and electrons (3 of each), and a handful of neutrons. Of course, adding or subtracting such particles to an element is easier said than done. Producing gold by fusing lighter elements is so difficult that gold was not formed in the furnace of early stars, despite their high energies. It was likely created in the thermonuclear reactions of colliding stars and blasted into the surrounding universe.[4] Tiny atoms can be very resistant to change.

You can build lots of structures with an assortment of Lego blocks. You could assemble blocks to look like a tree or to look like a dog, but the structures would remain just piles of blocks, with none of the properties of a tree or a dog. But add some atomic blocks to hydrogen and you get oxygen, and you can combine those two to get water, which can also turn into ice or steam. Amazing. It's as if you were building a shelter with sticks and then by adding a few more sticks it suddenly turned into a solid house!

Of course, scientists weren't content to settle with the theory of chemical elements being made of three fundamental particles and leave it at that. They wanted to know the properties of those particles and whether they could be divided even further. We might be interested, too.

Forces Within the Nucleus

One thing seemed strange right from the beginning: protons,

which are tightly compacted in the nucleus of every atom, are positively charged, and like charges repel each other. Something was keeping the protons (and the neighboring neutrons) from scattering apart.

This led to the discovery that protons and neutrons are not fundamental particles; they can be broken down still further into particles called quarks and their force-conveying partners, gluons. The latter, as the name suggests, exert a force at very short distances that keeps the nucleus together. Unlike some other forces, its strength increases with distance within the nucleus of an atom. The force is called the strong nuclear force, distinguishing it from the weak nuclear force, which has to do with the emission of radioactivity from certain atoms with "too many" protons.

Quarks have a number of strange properties, one of which is that they are never found alone. Current theory—collected in what's called the Standard Model of particle physics[5]—maintains that they are fundamental particles, no longer divisible into smaller parts.

Those who hope for simplicity in the underlying makeup of the world may have imagined that the number of fundamental particles would perhaps be small. If every chemical element in our world can be constructed from just three basic particles, what need is there of anything else? Trees and water, the sun and all the stars, us—we're all made of the same stuff.

But it was not to be. There is a whole Heinz 57 variety of fundamental particles out there. Some, like neutrinos, are so small that they fly through the Earth (and us) all the time without interacting. There are three kinds of neutrinos, all with no charge and possessing much less mass than electrons. They travel at nearly the speed of light. (A couple of years ago, scientists in Italy thought they had clocked neutrinos traveling faster than light, which would have violated Einstein's theory of special relativity and overturned a lot of physics, assuming that neutrinos do have mass. On further review, the neutrinos were found to be obeying the speed limit.)[6] Most of the neutrinos detected on Earth come from nuclear reactions in the sun.

There has been much success at grouping the many particles in the Standard Model according to their properties—like the periodic table of chemical elements—but it is a complicated picture, with physicists having to resort to some unusual names (such as "strange" particles and "charm" particles) in this veritable stew. Amazingly, however, scientists have been able to discern the laws that these particles follow. In fact, they obey them to a very high degree of accuracy, meaning that the Standard Model is a very reliable description of the fundamental building blocks of our universe.

However—there's always a However—the model is certainly incomplete. It seems to work fine in the ultra-small world of the atom, but does not fully incorporate gravity, which operates predominantly in the realm of stars and planets. There is speculation that gravity is the result of a force-conveying particle called a "graviton," but it has yet to be discovered. Moreover, astronomers who study the universe on the super-macro level have found that something more is holding galaxies together that is not accounted for by the Standard Model. And what's missing is not just a minor adjustment.

It appears that 75% of the matter in our universe is made up of something other than the particles we have previously identified. This missing stuff has been temporarily called "dark matter" for want of a more scientific description. Some theories attribute this missing matter to what are called WIMPs—weakly interacting massive particles. Weakly interacting because they are cold and dark, impervious to the electromagnetic force. Massive because they possess some mass and may be very heavy compared to other subatomic particles. Collectively they exert an enormous force on galaxies throughout the universe. Their massiveness may make them difficult to detect in powerful particle accelerators like the one at CERN in Switzerland, which was recently credited with discovering the Higgs boson (see below). To create massive particles with these machines you need a lot of energy. CERN is the most powerful accelerator in the world, but it may not be strong enough to create collisions resulting in something as heavy as dark matter. (Side note, the US intended to build an even more powerful collider in Texas before the present one at CERN was created. However, funding was cut and the US lost dominance in that field.)

Given the pace of scientific discovery, it seems likely that the nature of dark matter will be revealed in the near future. Whatever it is, its discovery will likely require a revision of the theories that explain not only the tapestry of the galaxies, but of the basic building blocks of the entire universe.

Revising current theories would not be a totally surprising outcome. Scientists have known for some time that Einstein's theory of general relativity, which is the current explanation for how gravity works, is not in line with the Standard Model of particle physics.[7] The equations go haywire at the small distances in the atom and with such tiny masses. Einstein had hoped to come up with a comprehensive theory that encompassed both the most massive and the tiniest objects, but he was not able to do so. Neither has anyone else, so far.

This large mix of fundamental particles, many of which seem to have little effect on our daily existence, offers yet another repudiation to the notion that we humans are the center and purpose of the universe. Just as our planet seems a tiny afterthought in a universe with billions times billions of other planets, so too, it seems the fundamental particles of our existence (protons, neutrons and electrons) are just a small part of the mix that nature has in store for other purposes.

The Higgs Boson

One of the most recent discoveries in the world of the very small is the verification of a particle called the Higgs boson. It is called Higgs after one of the scientists, Peter Higgs, who predicted its existence in the 1960s.[8] It is called a boson (as opposed to a fermion) because bosons include the particles that can convey force, and the Higgs was theorized as the conveyor of mass to other particles. But the discovery of the Higgs would probably not have captured the world's attention as prominently as it did in 2014 if it was only referred to as just the "Higgs boson."

Creative scientists realized that the funding needed to explore and find this elusive particle required creative marketing. Leon Lederman of the Fermilab in Chicago came up with the description of the Higgs as the "God Particle," a far more intriguing name. The hype has continued to the present day. Two of the excellent and readable accounts of the discovery of the Higgs particle evoke that same otherworldly quality, rescuing the Higgs from the obscure world of the Standard Model. Lisa Randall's book describing the physics and engineering leading up to the Higgs discovery is titled *Knocking on Heaven's Door*,[9] and Sean Carroll's account of the actual discovery is *The Particle at the End of the Universe*.[10]

Carroll admits that the Higgs was not hiding at the end of the universe:

As far as we know there isn't any "end" to the universe, either at some location in space or at some future moment in time. And if there were a location where the universe could be said to end, there's no reason to think you would find a particle there. And if you did, there's no reason to think it would be the Higgs boson.[11]

Nevertheless, he thought it was a fitting metaphor.

The Higgs particle has nothing to do with God or heaven, at least no more than any of the other pieces of the Standard Model puzzle. But, as Carroll explains, it has been a critical conundrum for a long time. With an ordinary puzzle, you can tell some features of a missing piece if you have completed the puzzle all around the space waiting to be filled. The missing piece should fit neatly within the rest of the puzzle, completing that part of a picture. Scientists similarly theorized that if the Higgs existed, it should have certain properties based on the properties of other particles that had already been discovered and formed an almost-complete puzzle. Not surprisingly, the puzzle is still not finished, even with the Higgs in place.

From the very beginning, scientists realized that converting the theoretical prediction of this addition to the Standard Model into an experimental result would take a huge commitment of resources and money. Subatomic particles are now typically found by using particle accelerators that smash known particles together at high energy in the hope that the remnants of the collision will reveal the fleeting existence of a new particle.

The predicted Higgs boson had not shown up in the existing particle detectors when it was theorized by Higgs himself in 1965. That meant that either the theory of its existence was wrong, or more powerful particle accelerators were needed to create more energetic collisions that would reveal larger unknown particles. More power requires more money—a lot more it turns out.

Less energetic collisions create less massive particles, following the principle that mass and energy are two forms of the same entity. The theory of the Higgs indicated it was a relatively massive particle and finding it would require simulating conditions similar to those existing a fraction of a second after the enormous expansion that resulted in the universe we live in today.

Over the next few decades it was hoped that accelerators such as the one at Fermilab or the earlier accelerator at CERN would be powerful enough to find the Higgs. However, no confirmation of a particle fitting the theoretical description was found.

When funding for the Superconducting Super Collider was

denied by the US Congress, a consortium of European countries (with support from the US) committed themselves to replacing the existing accelerator at CERN with a much more powerful machine. It would not reach the most energetic collisions envisioned for the Super Collider, but it was likely to reach energies higher than the outside limit at which the Higgs should be seen. In other words, it should be up to the job, and funds were committed to build it.

The story of the Higgs boson is as much a story of the daunting effort required to build the CERN accelerator and the exquisite experiments performed with it by a team of over 1,000 scientists from around the world, as it is a story about the particle itself. It may have been that challenge that helped CERN to be funded—nations working together, involving the brightest scientists making contributions towards a discovery that had eluded detection for 40 years. Even the prospect of finding that the Higgs didn't exist would be an important milestone in science. It would require that a new theory of the fundamental particles be formed.

Although the Higgs particle is not one of the basic elements of our life as we know it, it was theorized to serve a critical role in the atomic world. The particle was envisioned to be an excitation, a quantum, a piece of the Higgs field, which scientists believed permeates the universe. The field interacts with some of the fundamental particles and gives them their mass, that is, their resistance to move or change direction unless acted upon by some force.

Some particles, like the photons of light, do not interact with the Higgs field, and hence remain massless. Other particles, like electrons, do have mass, and it is the Higgs that determines it. One might have thought that some particles have mass because that's just the way they were "born." But the current theory of the earliest moments of the universe—the Big Bang—indicates that the tiny seed of our world was an undifferentiated cauldron of energy and that the particles we now recognize evolved in the first moments, gradually expanding, cooling, and interacting with each other. Some of those interactions gave certain particles a distinct mass. Those particles had to obey the atomic speed limit and travel below the speed of light, while massless photons of light could equal it.

The Higgs field slowed affected particles similar to the way a bowl of molasses would slow a marble dropped into it. The marble would behave as if it just became heavier and harder to move.

The accelerator at CERN has been called the largest and most complicated thing ever built by humans. It consists of a large underground oval, 17 miles in circumference. It passes through parts of Switzerland and France. Although the tunnels of this subway are huge, the travelers in it are among the tiniest passengers in the universe: protons. The protons are accelerated by a series of massive magnets, strategically placed around the oval until they reach a speed of 99.9% that of light. To make the forces involved even greater, another stream of protons is accelerated to the same speed but in the opposite direction around the track. At a critical juncture, the two streams of protons destructively collide head-on.

Many of the tiny particles in the opposite streams will fly by each other, but some collisions between protons will occur, creating a trail of debris that detectors can record and measure. It's in that confusing aftermath of the collision that the Higgs and other particles might briefly exist. Scientists would know that one of those particles was the Higgs if it was at an energy level that hadn't been seen in other particles, and if it predictably decayed into certain known subatomic particles.

The CERN accelerator had to be repaired after its first test run and that delayed the critical experiment for a couple years. The machine worked fine after the repairs and it was revved up to near maximum potential. Scientists observed that one output from the collisions indeed had an energy (or equivalently mass) of 125 million electron volts and appeared to decay into the expected sub-particles.

A different group of scientists, also working at CERN, but conducting another set of tests on the same streams of protons, also found their data pointing to the Higgs. All the evidence indicated a confirmation of the predicted results. No one could say they had seen the Higgs, since that would be impossible, even with this impressive machine. So, there is always the possibility that what has been discovered is very much like the predicted Higgs particle but is really something else that remains to be explained. But for now, the scientific community is convinced within a high degree of certainty that the Standard Model's missing piece has been detected and it fits neatly within the rest of the puzzle. Francois Englert of Belgium and Peter Higgs were awarded the Nobel Prize in Physics in 2013 "for the theoretical discovery of a mechanism that contributes to our understanding of the origin of mass of subatomic particles, and which recently was confirmed through the discovery of the predicted fundamental particle, by the ATLAS and CMS experiments at CERN's Large Hadron Collider."[12]

Chapter 9

Stranger Still

I think I can safely say that nobody understands quantum mechanics.

– Richard Feynman[1]

When it comes to explaining the discoveries of science in plain language, black holes were simple compared to quantum mechanics. Some of the most brilliant scientific minds of our time have concluded that nobody (including themselves) understands this micro world. Although Einstein was one of the discoverers of quantum strangeness, he spent much of his life trying to refute the implications that this discovery led to.

Nevertheless, after exploring the world of the very large — traveling from the big bang to the outer limits of the universe — it seems that we can't pass up delving into the fundamental building blocks of the universe. The quantum world is the world of elementary particles that don't act like particles at all. There's nothing "wrong" or even unusual with the way the universe behaves at this level. It is the just the way things are. The problem is that nothing in our evolutionary education has prepared us to understand the way these elementary entities interact, or even to know how to describe them.

In the macro view of the world, to build a universe you could start with a collection of hydrogen atoms. Let them be drawn together by gravity into a giant gaseous ball. As the ball's mass increases, so too will its pressure and energy, and the hydrogen atoms will travel at great speeds, violently colliding and fusing together. The fused atoms will yield helium, giving off an enormous amount of energy in the form of heat and radiation as a by-product.

This massive sun (star) will attract other matter into its orbit, and some of that matter will coalesce into smaller, cooler bodies around the sun. These planets may be rocky like the Earth, or gaseous like Jupiter. They, in turn, could have their own orbiting masses called moons.

Repeat many billions of times. That, oversimplified, is how the present universe came to be. It's a reasonable building plan. Gravity is the driving force pulling objects together on the macro level. On the micro level, one might think that since everything in our everyday world can be put together with just three particles (electrons, protons and neutrons), the world of the very small could be completely described using these building blocks.

However, it turns out that the universe contains many other fundamental particles that have little to do with us. Just as there are billions of galaxies with billions of stars that human life will probably never experience, so, too, there are many fundamental particles whose place and role in the universe is still a mystery. As indicated earlier, it is now believed that most of the matter in the universe has yet to be identified and has only been indirectly detected. We know it exists, but cannot identify what it is. Space is also permeated with energy in unusual forms, some of which we also cannot explain.

Particles may be the wrong word to describe the fundamental entities of the universe. There is a growing consensus that the building blocks of the universe are actually more like bundles of energy, agitated portions of invisible fields that fill all of space. Moreover, these quasi-particles have properties unlike anything else we've ever experienced. It's literally impossible to precisely measure their attributes—such as their exact position and speed at a particular moment. It's not just a question of the limitations of our measuring tools. These entities can be in many different places at the same time. When we attempt to measure one property accurately, another property then becomes impossible to measure with the same precision. Moreover, two different fundamental particles can be separated by hundreds of miles and yet maintain an instantaneous relationship with each other.

This may sound like science fiction. Our common notion of particles is that they behave like billiard balls. They are solid and move predictably. You can measure their properties as precisely as you like. For one ball to affect the other, they typically have to collide, and knowing the angle and speed of the collision will tell you what happens next.

However, billiard balls are not ideal models for fundamental particles. They occupy many points in space and can be broken down into more elementary substances. Describing the world at the atomic level escapes simple analogies. The building blocks are so small we can't see them, even with a powerful microscope. An elementary particle does not occupy a well-defined space. It can even possess two different characteristics simultaneously — not just being black *or* white, but all black *and* all white at the same time. Physicists call such entities "quanta." They are something like discrete pieces of energy, like a knot in a pine board — a concentrated disturbance in the wood grain, but still part of the wood. The way they behave and interact is called "quantum mechanics."

The characteristics of quanta are not contradictory or impossible to state. The problem is that describing them seems to violate our basic understanding of reality. In his book, *Tales of the Quantum*, physicist Art Hobson offered this vibrant but still challenging description of quanta:

[Q]uanta are highly unified, extended, flexible, changeable, unpredictable, fragile, gossamer entities that flit about or vibrate at mostly enormous speeds, that can extend over vast regions yet collapse in apparently a single instant of time to atomic or far smaller dimensions, that can vanish in an instant by transforming their energy into some other form, that can burst into being from the energy of a decaying atom or other quantum or from a random energy fluctuation, and that can entangle with each other to form highly unified yet delicately connected networks of composite correlated objects that appear to be in instant contact over perhaps cosmological distances.

Particle is exactly the wrong word to describe such an object.[2]

Shedding Light on the Subject

The mysteries of the quantum world began to appear as scientists probed the nature of light. Nothing could be more important to our existence than light. We depend on it to see, to eat, and to stay warm. One of the most familiar creation stories begins with the command: "Let there be light." But what exactly is light?

Isaac Newton carefully studied the complex properties of light. He passed light through a prism of glass and observed that a white beam could be broken into a rainbow of colors. In fact, rainbows themselves are created when light passes through the prisms of raindrops. He believed light was composed of tiny particles containing a small amount of mass. Because it has mass, Newton predicted that a beam of light would be bent by gravity as it came close to a massive body such as a star.

If, instead of a prism, you shine a concentrated beam of light through a narrow slit in a piece of metal and place a light-recording screen on the other side, a band of white will appear on the screen, as might be expected. Put two parallel slits in the metal barrier and you might expect to see two bands of white on the screen. But that is not what always happens. When the slits are sufficiently close together, what you see are numerous dark and white bands across the screen. Causing more intense light to go through the slits just produces more distinct bands, but the black patches in between do not go away. Light passing through the two slits appears to be broken into two waves that interfere with each other and produce a classic interference pattern of light and dark on the recording screen.

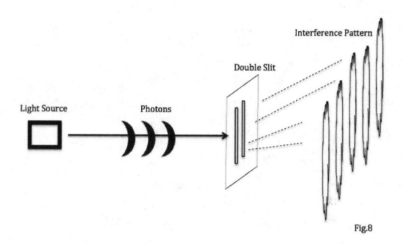

Fig.8

If light were simply a wave phenomenon, this interference would not be surprising. If you combine two water waves or two sound waves, they will also produce interference patterns. Waves from the movement of the ocean, for example, can encounter waves created by a passing boat. The result will be that some of the ocean waves appear to be strengthened, rising higher, and some seem to get flattened out. If the peak of one wave combines with the peak of another, their combined height will be higher than either one. And when a high point of one wave meets the trough of another, the whole wave may be flattened to zero, like the dark bands on the screen. Properly tuned, two sound waves can also cancel each other out, resulting in silence.

Recall from Chapter 2 that you can create a wave by moving a piece of rope up and down. If you have somebody take the other end of the rope, and similarly raise and lower her end of the rope, you will have competing waves interfering along the length of the rope. If you synchronize your motions—sensing the timing that produces a reinforcing rhythm—you will be able to magnify each other's actions, sending higher waves up and down the rope than would be the case if only one of you were waving.

Einstein's Insight

So far, so good. Light seems well behaved, exhibiting wave-like properties rather than the particle behavior that Newton had imagined. However, in 1905, Albert Einstein interjected a bit more mystery into the behavior of light.

In addition to working at his day job at the patent office in Bern, Switzerland, Einstein wrote three groundbreaking papers in his spare time, each of which were major contributions to science.[3] One of those papers, the one for which he eventually won the Nobel Prize, was about the photoelectric effect of light. From earlier scientific experiments, Einstein knew that when light strikes certain materials, electrons are released. If light behaved like a continuous wave of energy, then decreasing the energy of the light should still kick electrons out of the material, provided the light was directed at the material for a longer time. That didn't happen. Below a certain energy (that is, its frequency), the light did not free up any electrons.

Similarly, increasing the intensity (the brightness) of the light did nothing to increase the energy of the electrons being released. Changing the frequency (the color, but also the energy) of the light made all the difference. Einstein proposed that all of this made sense if light consisted of discrete packets of energy, called photons, and was not just a wave phenomenon. Photons of sufficient energy (frequency) could free up electrons in the material. Higher frequency photons could impart even more energy to the impacted electrons.

Einstein's proposal that light was not simply a continuous wave echoed earlier discoveries by Max Planck in 1900.[4] Planck found that when you increasingly heat a metal surface, it does not radiate in a continuously increasing manner, but instead has quantum jumps in its radiation. Both visible light and radiation from a heated surface are parts of the same phenomenon: electromagnetic radiation that exhibits wave- and particle-like properties and travels at the speed of light.

The weirdness of the quantum world comes in when you harken back to the two-slit experiment with light. A wave going through two slits might produce interfering waves on the other side of the barrier. But experimenters have been able to slow the frequency of light down sufficiently so that only individual photons (packets of radiation) are sent towards the two slits. How do discrete packets of light produce wave-like interference patterns? As each photon makes its way past the slits (presumably passing through one slit or the other), it makes a single exposure on the recording screen. When repeated many times, you might expect to see two bands on the screen, one for each slit.

But as indicated earlier, that is not what happens, even when individual photons are sent through the slits. Instead, after numerous photons have passed through, the interference pattern of multiple light and dark bands appears. The bizarre explanation of quantum mechanics is that each whole photon (they are indivisible) passes through *both* slits at the same time and interferes with itself!

If you think of the packets of light as billiard balls, the image makes no sense. A whole ball can't go through the left slit and simultaneously go through the right slit. But photons accomplish this trick. After a photon passes through both slits, it hits the receptor screen, creating only one impression. But the sum of a large series of impressions produces the interference pattern of a wave. It doesn't seem possible to us, but that is the world of quantum mechanics and the world we live in.

Fields of Light

Scientists later concluded that visible light was in fact just one part of the broad spectrum of electromagnetic radiation. The only difference between light that we can see and radio waves or microwaves that we can't is the frequency of the waves, that is, how many times per second they rise and fall. All electromagnetic waves travel at the same speed—the speed of light—through the electromagnetic field.

What are fields?[5] They were touched on in Chapter 2 when discussing gravitational waves. A field is a designated space in which every location has an assigned value. Fields are all around us and throughout space. Most people are comfortable with the notion of water waves moving through the field of the ocean. Light traveling through the electromagnetic field seems more mysterious, but the principles are the same. If you could take hold of the electromagnetic field and wave it like a blanket at the beach, you could create electromagnetic waves that would travel through our solar system and beyond. An alien with a TV antenna might be able to see the pattern of your wave.

Of course, there are no handles on the electromagnetic field, but every time you turn on a light you send a small wave through this field. Another common example is talking into a radio microphone. The sound waves of your voice travel through the field of air and cause a diaphragm in the microphone to move. The diaphragm is connected to an electric circuit, which converts the vibrations into an electrical energy signal, similar to the way the short and long pulses of Morse Code can be sent along an electric wire. For wireless transmission, the electronic signal can then be amplified and sent into space as electromagnetic waves from an antenna. When another antenna is impacted by that signal, the whole process is carried out in reverse: the electromagnetic waves become an electric signal, which moves a diaphragm in a speaker, creating sound waves that you can hear on your radio. Television operates in a similar way.

Rather than saying that light behaves like a wave at times and like a particle at other times, it is probably more accurate to say that light is a phenomenon that does not fit into our usual human categories. Light—at its fundamental level—is like nothing we've experienced before. We recognize some of its behaviors, but its actual form is strange to us. Eventually, we will get used to its uniqueness.

That day should come quickly because the quantum world has more surprises for us. It turns out that the other elementary particles—like electrons—behave like photons of light. They are not like billiard balls, solid and predictable. Instead, they sometimes behave like particles and sometimes like waves. A beam of electrons going through two slits produces an interference pattern, just as light waves do. The same can be said of protons, though they are much more massive than electrons and it takes more energy to see this effect.

The Inherent Uncertainty of the Quantum World

Physicists have concluded that much of the quantum world is unknowable in ordinary terms. Early images of the atom pictured electrons as tiny particles orbiting the nucleus, like a planet orbiting the sun. Instead, they are more accurately described as excitations within a vast field and hence cannot be assigned to any one place in space at a given time.

It is not possible to precisely predict the location of an electron in, say, a hydrogen atom, even though a measurement will return a definite location. Other measurements on other hydrogen atoms will return different answers. In a sense, the act of measuring forces the electron to "choose" a single location from the many it simultaneously occupies. The undefined and amorphous dispersal of the electron collapses into a point whenever a measurement is taken of its location.

Some locations of the electron are more likely to show up than others. Before measuring, there is a correspondence between potential locations and a probability that the electron will be there when measured. For example, perhaps there is a 75% probability that an electron will be "found" in the innermost shell around the nucleus of an atom, but there is a 25% probability it will be elsewhere. If you measure the electron's position many times, it is very likely that 75% of your results will show the electron in the innermost shell, just like if you flip a coin enough times, half of the results will be heads.

Particle physicists have actually come up with a very precise equation (thank you, Erwin Schrödinger)[6] that specifies the probability for any location of the electron. If you stick a location into the formula, it can spit out the probability of finding the electron there. However, until measured, it is seemingly everywhere at the same time. Einstein objected to the notion that electrons had no particular location until measured. He said, "God does not play dice." Instead he believed that there were hidden variables that, if known, would reveal where an electron would actually be found.

Electrons that are part of an atom are most likely to be relatively close to the nucleus of that atom, as opposed to being located way out in space. Moreover, electrons have preferred shells at discrete distances from the nucleus. Picture the various running lanes of an oval track. You might have one electron at the innermost lane, two in the next outer lane, etc. Until measured or observed, the electron has some probability of being almost anywhere, but a series of measurements will reveal that it is far more likely to be in one of its usual shells.

If you take two atoms, both of their electrons are most likely to be in their predicted shells. The odds of both atoms having electrons outside of those shells are even smaller than for a single atom. Hence, when you take an ordinary object, which is made of billions of atoms, the odds are astronomically great that the object will have a normal appearance and not be drifting from one location to another. That is the everyday world we observe with our eyes. When we look up at the moon, it is in one place at a given time, and we can predict with almost certainty where it will be tomorrow and ten years from now. (Hence, the success of predicting an eclipse, or of landing on the moon itself.)

From the human perspective, the imprecision of the quantum world is actually worse. If you want to measure two related aspects of a fundamental particle, say its location and its momentum (the product of its mass and velocity), there is a distinct margin of error in the combined results that cannot be overcome. Either you can pin its location down to a very specific number and have to settle for a large uncertainty in determining the particle's momentum, or you can pin down its momentum and be very unsure of its location. This limitation on observations of the quantum world is called the Heisenberg Uncertainty Principle and was stated by Werner Heisenberg in 1927.[7]

The problem is not that the tools for measuring are imprecise or even that measuring something as small as an electron is bound to be impacted by the method you use to measure it (say, shining a light upon it). Rather, the uncertainty arises because fundamental particles do not have specific locations and speeds. They have wave-like properties and have the characteristic of being in many different states at the same time. Again, this doesn't make sense to our way of thinking, but that doesn't change the reality.

With elementary particles, it is not just their location or momentum that defies precision. Other qualities also only exist in a suspended state of uncertainty. For example, an electron has an attribute similar to the spin of a top. It is said to either spin up or spin down, similar to clockwise or counterclockwise spin. But until you attempt to measure which way the spin is pointing, the electron is spinning up AND down. An ordinary spinning object is either rotating clockwise or counterclockwise, not two directions at the same time. In fact, just as an electron has some chance of being found far away from an atom's nucleus, there is some chance that the electron's spin is pointing somewhere between up or down, as well.

Actual tops only spin one way, even though they are composed of electrons (and neutrons and protons) capable of spinning many different ways at the same time. It's not just that spinning in two directions at once is hard to accomplish, it's a phenomenon completely outside the meaning of our words. Probability is what saves us from going crazy when viewing ordinary objects like tops. Each electron in the top has an equal probability of spinning up or down, and with billions upon billions of electrons, the different directions cancel each other out and have no effect on the ordinary spinning of the top, which we can control with a twist of our hands.

What Else is Weird? Spooky Action at a Distance

Imagine if two people—scientists almost always call them Alice and Bob—are separated as far apart as you like, and each has a magic penny. Each penny has a head and a tail, but they magically act like one coin. If Alice flips her coin and it comes up heads, then Bob's flip will be tails, and vice versa. How does Bob's coin "know" what to show so as to complement Alice's toss? There's no time for a message to be sent between them, the result occurs instantaneously (though it would take some time for Alice and Bob to communicate their results to each other).

Science abhors magic. What you expect to see in this experiment is that the coins act independently. Whatever the result of Alice's flip, Bob's coin should have a 50% chance of being heads and a 50% chance of being tails. What is exercising this control at distances that can even be light-years away?

Even though these coins violate our common sense notion of location (separate objects act independently unless there is a connection between the two), fundamental particles can act just like the magic coins. It's called entanglement, and it is common at the atomic level. The difference between the coins and the electrons is that there is nothing magical about the particles. The electrons of the entangled pair are just like every other electron, except for their mutual relationship, which is relatively easy to produce.

If two complementary electrons are entangled (usually that means produced from the same source), and each can randomly have an up or down spin, then if you measure the spin of one of these electrons, the other will have the opposite spin, even when located far from the other. One electron seems to instantaneously "know" what direction the other electron was observed at, and it instantly "knows" to produce the opposite result every time. This is not just a theory; it is observed regularly in experiments.

This is called action at a distance, or "spooky" action at a distance according to Einstein.[8] He did not like this notion that distant particles could somehow instantaneously influence each other. Newton's theory of gravity, which Einstein's theory supplanted, suffered from the same mysterious defect: gravity had to instantaneously operate between two distant objects with a force Newton could not explain. Einstein's theory of gravity cured that problem by having gravitational waves travel at the speed of light (fast, but not instantaneous) and displaying itself as the bending of space itself, thereby describing how massive objects were affected by the presence of each other. Einstein believed that perhaps there was also a deeper law that accounted for the instantaneous action at a distance for fundamental particles.

In sum, the quantum world is unlike anything else we've experienced. Fundamental particles do not exist in one place, and two (or more) of them can instantaneously interact with each other, despite lacking any means of communication. Their location is undefined; they are energetic chunks of an invisible field. The seeming solidity of our everyday world is an illusion. Solids are mostly empty space, and even at a level of fundamental particles, their hardness comes not from impenetrable cores but from force fields that repel each other — something seemingly out of a *Star Wars* battle station. And yet, these elementary pieces, these quanta, are not just arcane entities that theoretical physicists imagine being out in space; they make up the chemicals and molecules that are our bodies and everything we see around us.

Chapter 10

On The Horizon

Can it really be that at its most fundamental level the universe is divided, requiring one set of laws when things are large, and a different, incompatible set when things are small?
– Brian Greene[1]

New developments in science are emerging every day. It's impossible to keep up with them all. This chapter briefly touches on two areas that are not entirely new, but are still at the frontiers of science and that would have enormous impact if the theories conform with reality. The first area of research is a practical one: the building of a working quantum computer, which has recently moved from a theoretical idea of the 1950s to a working startup at a few cutting-edge laboratories around the world.

The second area is the much more theoretical world of string theory, which has been mentioned in earlier chapters. It attempts to explain fundamental particles and forces in a whole new way. It was proposed over twenty years ago, but is constantly evolving and remains untested. Its importance lies in the promise that if true, it could prove to be the Holy Grail of physics—a theory of everything.

Quantum Computers

When I was a freshman in college, those studying science carried slide rules. These manual calculators were one step up from an abacus but worked well in giving approximate answers to messy calculations in chemistry and physics. True nerds had fancy slide rules in cases attached to their belts.

In the atrium of the college mathematics building was a giant computer. You could watch it printing out intricate designs according to instructions fed in with punch cards. In my four years majoring in mathematics, I never had occasion to use or even visit that, or any other, computer. Of course, no one had a laptop or a computer in their dorm room. PCs did not exist. Few realized how dramatically computers would change the world.

Today, any smart phone has vastly more computing power than that whole room of electronics in the math building. The growth of the computer industry has paralleled something called Moore's Law, which predicts that the number of transistors (tiny electronic switches) that can be fit on a computer chip will double every 18 to 24 months. Gordon Moore, one of the founders of Intel—a leading computer-chip company—made the prediction in 1965, originally forecasting a doubling every year for the coming decade.

Doubling is a deceptively fast-growth process. If you start with the number 2 and double it, you get 4, and then 8, 16, etc. Those are not impressively large numbers, but if you allow that doubling to continue, you soon arrive at numbers of immense size. For example, 2^{32} is equivalent to 2 used as a multiplier 32 times, or precisely 4,294,967,296. This demonstrates the power of doubling, or exponential growth, and makes for some pretty crowded computer chips as the number of switches squeezed onto a given area repeatedly doubled over forty years.

The law—which was really more of an educated guess— has proven remarkably predictive. A graph of the number of transistors on a chip over time shows a steady doubling over forty years from 1971 with 2,300 transistors to over two billion per chip in 2011, when the growth rate started slowing.[2]

Computer switches are made from ordinary chemicals like silicon and copper and ultimately from combinations of atoms, which in turn are made from electrons, protons and neutrons. As manufacturers kept shrinking the size of the switches in their computers, they have approached the size of a few atoms. Those atoms are made of fundamental particles that exhibit strange behaviors, as discussed in Chapter 9. Using switches close in size to the atomic level could interject a high degree of unpredictability into a circuit, which might not be the best recipe for computer accuracy.

But rather than being a barrier to more powerful computers, scientists see the world of atomic building blocks as the wave of the future. Research is underway around the world to develop a working quantum computer, one that takes advantage of the strange behavior of elementary particles.

Fingers and Toes

Current computers use a very simple language—the language of two. Information must pass through electrical switches, and the switches are either ON or OFF, which can be represented by the numbers 1 or 0. One and zero are the basis of the binary number system. Punch cards were one of the earliest ways of inputting information into computers in the binary system because each hole (or the absence of a hole) in a particular location on the card could tell a switch to be either on or off, a one or zero.

So, where is the 2 in this two-digit system, you might ask? In the two-digit number system, 2 is 10 (read "one, zero"), just as the number ten in our ten-digit number system is also 10. It's not an additional digit; it's a combination of two smaller digits, where the placement of each matters. When you have used up the number of different digits in your number system, you introduce a two-place number and start counting with your digits all over again. We count from 0 to 9 using ten single digits. Then we introduce a two-place number, 10, and we can count up to 19 by gradually increasing the last (to the right) digit: 11, 12, 13, etc., and we can count up to 99 by also increasing the first (left) digit. To put it another way, forty-three is 4 in the tens column plus 3 in the units column, or 43.

The reason we have a ten-digit number system is because we have ten fingers, the original means of counting, both among children and ancient civilizations. There's nothing special about a ten-digit system, especially to a machine. If we had six fingers like the three-toed sloth, we probably would have devised a six-digit number system—using 0 to 5. (The advantage of starting with zero in a number system was not immediately obvious and has an interesting history in mathematics.)[3] The number six would be 10 because we'd run out of digits after 5. It would represent 1 six in the sixes column and 0 in the units column. Similarly, 11 ("one, one") would be 7.

Any whole number can be expressed in the binary system. Just as in the ten-digit system, the number of columns or places indicates how many multiples of two the first (reading left to right) digit represents. The binary 10000 has five places. The last digit (right end) is the units place, that is, no multiples of two. The next place to the left is multiples of two, the next is multiples of four, etc. So 10000 indicates one multiple of 16 and zero multiples of the other places, or exactly 16. Similarly, 111 in the binary system represents one *four*, one *two*, and one *one*, which is equivalent to 4+2+1=7.

Machines can handle the conversion process from ordinary numbers to binary, so there's no need to become familiar with the equivalent expressions. Letters of the alphabet are easy to convey, too, as soon as each letter is assigned a number, and the computer is given a signal that a string of letters (rather than numbers) is coming. Word processors work on that principle. Pictures nowadays are usually made of pixels, fluorescent dots on a screen. Pixels can be turned on or off, creating an image. Using three pixels—one for each of the primary colors—all the colors of the visible spectrum can be digitized, copied, and reproduced flawlessly. Videos can be similarly represented. All the information in the world can be put in a very long string of 1s and 0s.

Putting Quantum Weirdness to Work

With so much information coming in long strings of 1s and 0s, you can see why it's important for a fast computer to have a lot of capacity for receiving and processing data in a small space. Quantum computers accomplish this by using atoms or subatomic particles as the initial receptors of information. Such particles can be set in different states, say spinning up or down, just like switches are either on or off. But atoms are not just better components because they are smaller than conventional materials. Each subatomic particle has quantum properties, like being in more than one state at the same time and being able to work in harmony with another entangled subatomic particle (remember Alice and Bob's magic pennies) that can be utilized to speed up computing exponentially.

Quantum computers, like quantum mechanics generally, are complicated. The whole field is in its infancy, similar to classical computing in the early part of the 20th century. But the basic idea is that atoms can act as the holders and processors of information received by a computer. However, unlike an ordinary switch, which is either on OR off, a quantum switch can be on AND off at the same time. A single quantum bit (qubit, for short) can represent 1 AND 0 simultaneously.

To see how this helps in dealing with incoming numbers, consider a set of three classical computer switches.[4] Each switch can be either on or off. Together, they can be configured in eight different combinations. They can represent any of the eight numbers from 0 to 7, but only one such number at a time. For example, 5 would be 101 (on, off, on). If you wanted to insert all eight numbers into the computer, you would need eight configurations of these three-switches.

However, just one configuration of three quantum switches could handle the same task. Unlikely as it may seem, every number between 0 and 7 could be loaded into the same three atomic switches at once. Each atomic switch can represent "on" and "off," holding both in superposition. Viewed in a row, the switch furthest to the right (Switch 1 below) has that capacity. It can be part of representing 5 (on, off, ON); it can also be part of representing 4 (on, off, OFF) at the same time. And since each of the other switches has that same capacity, eight different numbers can be held in superposition by just three quantum switches.

Classical computer: two sets of switches to represent 5 and 4

Switch 3	Switch 2	Switch 1	
On	Off	On	= 5
On	Off	Off	= 4

Quantum computer: one set of switches—superposition in Switch 1

Switch 3	Switch 2	Switch 1	
On	Off	On and Off	= 5 and 4

The important thing to realize is that the representation of the two numbers above is accomplished by ONE quantum computer configuration. And the other six numbers can all be incorporated into this same configuration by utilizing superposition in Switches 2 and 3.

It is not hard to see how quickly a quantum computer could overtake a regular computer in sheer efficiency. Instead of eight sets with three switches each (24 switch configurations in all) to represent eight numbers, because of superposition, only three quantum switches are needed: $2 \times 2 \times 2 = 8$. This can be generalized: n quantum switches can represent 2^n different numbers, $2 \times 2 \times 2 \ldots \times 2$, with n multiples of 2. A very large quantity of information fed in as numbers could be handled by a small number of switches, as compared to an ordinary computer switch that can represent only one bit (0 or 1) at a time. To appreciate the potential of quantum computers for large groups of numbers, recall the power of doubling. Thirty-two (32) quantum switches could handle the same work as 2^{32} or 4,294,967,296 normal switches.

There are, however, some hurdles to clear. Recall that when a particle such as an electron or photon is measured or observed, the superposition of its many states collapses into one state. It's as if the particle randomly chooses one state whenever a measurement or observation is performed. The duality of being on AND off reverts to being just on OR just off, which relegates it back to an ordinary computer switch. If you peeked inside a quantum computer and looked at the string of three atoms simultaneously holding the digits 0 to 7, each atom in the string would randomly select on or off, and you would see only one of the eight digits represented in the string, say 5, and the computer would stick with it. So don't peek.

The trick is to keep the atoms in a suspended state until they are finished with their computing. In some of the experimental quantum computers, such a system has to be held at a temperature colder than outer space. So far, quantum computers have only been able to maintain the superposition state for a few seconds.[5] With their greater capacity to calculate, that would be fast enough for a few powerful computations, but not nearly long enough to churn out the huge calculations that regular supercomputers are capable of.

Another of the curiosities of particles in the quantum world may come into play here. Two subatomic particles can be "entangled," that is, despite being physically apart, they can have complementary outcomes when they are observed. If one entangled particle is observed and it randomly assumes a single state of spinning up, the other entangled particle will automatically choose spinning down. Imagine two dancers performing a perfectly synchronized routine, except that instead of dancing together, they are in different rooms and there is no music playing and no starting time assigned. This provides a quantum computer with a way of error correction. Two different sets of information can be known by only examining one, if the sets are entangled.

There is a bit (sorry) more to quantum computers than just keeping the computer shielded from outside observation. The rules that enable an ordinary computer to calculate are handled by "logic gates" (such as the configuration that produces an outcome of 1 only if BOTH incoming circuits read 1; all other combinations produce an outcome of 0; this is one of the gates needed for adding numbers). The logic gates for quantum computers are different from those in classical computers.[6] Quantum computers not only process faster than classical computers, they also logically compute in a different manner. Quantum computers will be able to handle problems considered unsolvable today, and they will do complex calculations in a much shorter time than ordinary computers.

Classical computers have already reached practical limits as some everyday problems would require a computer larger than the entire Earth, or would take longer than the history of the universe to solve. Consider the complexity of the human body with its billions of cells and trillions of neurons, all interacting. Even weather prediction is subject to the "butterfly effect": one wisp of air affects the local environment, which in turn impacts, in some small way, the weather hundreds of miles away.

Because science has persisted in understanding the world at the atomic level, and because that level has been found to be unlike anything we expected, we are now on the cusp of inventing a new wheel that will change civilization in ways we can't predict. At a minimum, the coming of quantum computers may help us understand and appreciate the quantum world that is so foreign to our common experience, but an integral part of who we are.

String Theory

The Standard Model of particle physics works well in the nuclear realm and is able to accurately describe how fundamental particles interact. Einstein's theory of general relativity and gravity, on the other hand, has contributed enormously to our understanding of the stars and galaxies. But as Brian Greene skillfully explains in *The Elegant Universe*,[7] Einstein's theories do not account for the strange unpredictability of quantum mechanics, and quantum mechanics does not offer a clear role for the influence of gravity.

Viewed as a force of attraction, gravity is by far the weakest of the four fundamental forces in nature. It is trillions of trillions of times weaker than the Weak Nuclear Force, and weaker still compared to the Strong Nuclear Force and the Electromagnetic Force.[8] There is speculation that gravity is associated with a fundamental particle, the graviton, but it has yet to be experimentally discovered.

You know you have a challenging problem if one of the greatest scientists who ever lived was unable to solve it in a lifetime of work. After his breakthrough theories about time and space became widely acclaimed, Einstein spent much of the rest of his life trying to devise a single theory incorporating both the quantum world and the gravitational world. He was convinced there must be one grand theory that governs both the very small and the very large parts of the universe.

The immensity of that problem helps explain the widespread interest in String Theory.[9] It provides a framework for understanding both of these divergent realms and incorporates ideas about mysteries such as dark matter and black holes. The basic idea is that the fundamental building blocks of the universe are not quantum particles but are rather tiny one-dimensional strings. One string is differentiated from another by the way it vibrates, similar to the way different musical notes can be produced by vibrating simple strings of varying thickness.

As the theory is scaled up to our more familiar atomic world, string theory predicts the particles and forces of the quantum world, such as electrons and photons. Gravity also flows out naturally from the theory as the product of a particular vibration of strings. The strings themselves are so small that they can probably never be observed by human instrumentation. The scale of measurement is so far outside of our current technology that it is difficult to even conceive of a way of conducting verifiable tests.

While string theory is mathematically appealing and offers a grand unification of science, it has also produced skeptics. For example, one interpretation of string theory proposes a universe with ten spatial dimensions. Three dimensions are visible to us, and the rest are curled up so tightly that only gravity operates within them. Adding one dimension that we cannot see and have never experienced is quite a leap. A ten-dimensional world can seem even more absurd. Other dimensions are an interesting idea, but there is no evidence that they exist.

As string theory has progressed, it has broadened to include many versions of the original theory, depending on which initial conditions are assigned to the prospective universe. Skeptical eyebrows rose to new heights when it was announced that there might be as many as 10^{500} allowable string theories, each predicting a different outcome for the universe. A theory is useful if it provides a way of predicting how things got to the way they are now, and how they will continue to develop in the future. If the set of theories that constitute string theory predict almost any conceivable universe, then it may not be very valuable as a scientific tool.

Still, some of the brightest minds in physics and cosmology have been attracted to the power and ultimate simplicity of this theory. Time will tell whether Einstein's dream of a theory of everything that unites all the forces of the universe into one equation will be realized through string theory. Stay tuned.

Notes

Introduction

1. Hawking, Stephen and Leonard Mlodinow, *The Grand Design*, Bantam Books 2010, p. 161.

Chapter 1

1. Livio, Mario, *The Accelerating Universe: Infinite Expansion, the Cosmological Constant, and the Beauty of the Cosmos*, John Wiley & Sons, Inc. 2000, p. 264.

2. Speed = distance/time. The distance in this calculation was actually the change in distance between Jupiter and Earth from when they were close compared to when they were further apart. Similarly, the time was the additional time for Io to appear if Jupiter was farther from Earth compared to when it was closer. See <https://www.amnh.org/explore/resource-collections/cosmic-horizons/profile-ole-roemer-and-the-speed-of-light/> (Accessed August 8, 2018).

3. Again, speed = distance/time. In this case, distance corresponds to the wavelength of light or twice the distance between melted spots in the cheese, and the time is the fraction of a second for one wavelength to pass, obtained from the labeled frequency of the microwave oven. See <https://www.youtube.com/watch?v=Fhu15X56acU> (Accessed August 8, 2018).

4. Singh, Simon, *Big Bang: The Origin of the Universe*, Harper Perennial 2004, pp. 178–85.

5. Elizabeth Howell, "How Many Galaxies Are There?" Space.com, March 19, 2018, at <https://www.space.com/25303-how-many-galaxies-are-in-the-universe.html> (Accessed August 8, 2018).

6. Adams, Douglas, *The Hitchhiker's Guide to the Galaxy*, Pan Books 1979.

7. See Greene, Brian, *The Elegant Universe: Superstrings, Hidden Dimensions, and the Quest for the Ultimate Reality*, Vintage Books 2000, pp. 81–82.

8. See e.g., Barrow, John D., *The Book of Nothing: Vacuums, Voids, and the Latest Ideas about the Origins of the Universe*, Pantheon Books 2000.

9. For a discussion of this discovery, see Carroll, Sean, *From Eternity to Here: The Quest for the Ultimate Theory of Time*, Dutton 2010, pp. 55–57; see also Livio, Mario, *The Accelerating Universe: Infinite Expansion, the Cosmological Constant, and the Beauty of the Cosmos*, John Wiley & Sons, Inc. 2000.

10. See Greene, Brian, *The Hidden Reality: Parallel Universes and the Deep Laws of the Cosmos*, Alfred A. Knopf 2011, pp. 24–25.

11. See Randall, Lisa, *Knocking on Heaven's Door: How Physics and Scientific Thinking Illuminate the Universe and the Modern World*, HarperCollins Publishers 2011, pp. 119–20.

12. Ferris, Timothy, *The Whole Shebang: A State-of-the-Universe(s) Report*, Simon & Schuster 1997.

13. See Vanessa Thomas, "Dark energy and the accelerating universe," *Astronomy* magazine, July 2002; see also Greene, Brian, *The Fabric of the Cosmos: Space, Time, and the Texture of Reality*, Alfred A. Knopf 2004, p. 299.

14. Brian Greene describes the size of the early universe as a "Planck-sized nugget." *The Elegant Universe*, Note 7 above, p. 358. The Planck length is 10^{-33} centimeter.

15. In *A Brief History of Time: From the Big Bang to Black Holes*, Bantam Dell Publishing 1988, Stephen Hawking discusses the theory of black holes and how, despite their name, they can give off radiation, thereby slowly evaporating. For a typical black hole, this process would take 10^{67} years, far, far longer than the current age of the universe. See also John Barrow and Joseph Silk, "How Will the World End?" in Ferris, Timothy, ed., *The World Treasury of Physics, Astronomy, and Mathematics*, Little, Brown and Company 1991, p. 426.

16. See Greene, *The Elegant Universe*, Note 7 above, p. 346 and subsequent discussion.

17. See Kaku, Michio, *Einstein's Cosmos: How Albert Einstein's Vision Transformed Our Understanding of Space and Time*, Atlas Books 2004, especially Chapter 3.

Chapter 2

1. Ferris, Timothy, *The Whole Shebang: A State-of-the-Universe(s) Report*, Simon & Schuster 1997, p. 22.

2. See generally Susskind, Leonard, *The Black Hole War: My Battle With Stephen Hawking to Make The World Safe for Quantum Mechanics*, Little, Brown and Company 2008, for a discussion of the more technical aspects of black holes.

3. See, e.g., <https://en.wikipedia.org/wiki/Black_star_(semi classical_gravity)> (Accessed August 8, 2018).

4. See, e.g., Randall, Lisa, *Knocking on Heaven's Door: How Physics and Scientific Thinking Illuminate the Universe and the Modern World*, HarperCollins Publishers 2011, Chapter 10, discussing the fears that experiments at the Large Hadron Collider in Geneva could produce miniature black holes that would swallow everything around them.

5. See Greene, Brian, *The Hidden Reality: Parallel Universes and the Deep Laws of the Cosmos*, Alfred A. Knopf 2011, pp. 240–41.

6. See <https://en.wikipedia.org/wiki/Nuclear_fusion> for a thorough discussion of this process (Accessed August 8, 2018).

7. John Wenz, "Astronomers Discover the Brightest Early Galaxy Ever," *Astronomy* magazine, July 9, 2018 <http://astronomy.com/news/2018/07/astronomers-discover-the-brightest-early-galaxy-ever> (Accessed August 8, 2018). The black hole is at the center of a galaxy.

8. Susskind, *The Black Hole War*, Note 2 above; see also Greene, *The Hidden Reality*, Note 5 above, Chapter 9.

9. Hawking, Stephen et al., *The Future of Spacetime*, WW Norton & Co. 2002, pp. 87–108.

10. For a discussion of the pervasiveness of fields at the most fundamental levels of science, see Brooks, Rodney A., *Fields of Color: The theory that escaped Einstein*, Epic Publications 2010.

11. For a description of this groundbreaking experiment, see Dennis Overbye, "Gravitational Waves Detected, Confirming Einstein's Theory," *New York Times*, February 11, 2016.

12. Kenneth Chang, "It Was a Universe-Shaking Announcement. But What Is a Neutron Star Anyway," *New York Times*, October 16, 2017.

13. See Kip Thorne, "Spacetime Warps and the Quantum World: Speculations About the Future," in Hawking, Stephen et al., *The Future of Spacetime*, WW Norton & Co. 2002, p. 138.

Chapter 3

1. Quoted in Dennis Overbye, "A New View Of Our Universe: Only One of Many," *New York Times*, October 29, 2002, D4.

2. John Wenz, "Astronomers Discover the Brightest Early Galaxy Ever," *Astronomy* magazine, July 9, 2018 <http://astronomy.com/news/2018/07/astronomers-discover-the-brightest-early-galaxy-ever>, noting the earliest galaxy (Accessed August 8, 2018).

3. See "Observable universe," <https://en.wikipedia.org/wiki/Observable_universe> (Accessed August 8, 2018).

4. This has been referred to as the "horizon problem" because wherever in the universe one looks, the temperature of the cosmic background radiation is about the same. See Greene, Brian, *The Elegant Universe: Superstrings, Hidden Dimensions, and the Quest for the Ultimate Reality*, Vintage Books 2000, p. 353.

5. Mara Johnson-Groh, "Inflation leaves its mark," *Astronomy* magazine, August 2018, pp. 33–35. The faster growth of the

universe during the inflationary period allows more time in the pre-inflationary time for the particles to intermingle. See Hawking, Stephen and Leonard Mlodinow, *The Grand Design*, Bantam Books 2010, p. 130. This is also discussed in the next paragraph of the main text.

6. See Ferris, Timothy, *The Whole Shebang: A State-of-the-Universe(s) Report*, Simon & Schuster 1997, Chapter 9.

7. See Johnson-Groh, Note 5 above, p. 35, also indicating some doubts about the theory.

8. See Siegfried, Tom, *Strange Matters: Undiscovered Ideas at the Frontiers of Space and Time*, Joseph Henry Press 2002, pp. 122–23.

9. Dennis Overbye, "Astronomers Hedge on Big Bang Detection Claim," *New York Times*, June 19, 2014.

10. Greene, Brian, *The Hidden Reality: Parallel Universes and the Deep Laws of the Cosmos*, Alfred A. Knopf 2011, pp. 53–56.

11. For a discussion of Einstein's use of non-Euclidean geometry, see Kaku, Michio, *Einstein's Cosmos: How Albert Einstein's Vision Transformed Our Understanding of Space and Time*, Atlas Books 2004, pp. 92–102.

12. Abbott, Edwin, *Flatland: A Romance of Many Dimensions*, Seeley & Co. 1884.

13. See Greene, *The Elegant Universe*, Note 4 above, Chapter 8. Greene uses the analogy of an ant on a hose.

14. Lindley, David, *Where Does the Weirdness Go? Why Quantum Mechanics Is Strange But Not As Strange As You Think*, Basic Books 1996, pp. 107–11.

Chapter 4

1. Quoted in Charles Siebert, "The Genesis Project," *New York Times Magazine*, September 26, 2004, p. 55.

2. Ross Pomeroy, "Will Hitler Be the First Person That Aliens See?" RealClear Science, September 19, 2013 <https://www.realclearscience.com/blog/2013/09/will-hitler-be-the-first-

person-that-aliens-see.html> (Accessed August 8, 2018).

3. Michael Bakich, "Voyager's Grand Tour," *Astronomy* magazine, August 2018, pp. 30–32.

4. Kenneth Chang, "Falcon Heavy, in a Roar of Thunder, Carries SpaceX's Ambition Into Orbit," *New York Times*, February 6, 2018.

5. John Wenz, "The Red Planet revealed," *Astronomy* magazine, August 2018, p. 26.

6. Carolyn Porco, "Cassini at Saturn," *Scientific American*, October 2017, pp. 78–85, especially p. 84.

7. SETI.org at <https://seti.org/about-us/mission> (Accessed August 8, 2018).

8. See Michael Carroll, "The hunt for Earth's bigger cousins," *Astronomy* magazine, April 2017, pp. 22–27.

9. Liz Kruesi, "Top 10 Space Stories of 2017," *Astronomy* magazine, pp. 20–29. The Trappist story is on p. 28, second among the top stories.

Chapter 5

1. Carroll, Sean, *From Eternity to Here: The Quest for the Ultimate Theory of Time*, Dutton 2010, p. 13.

2. Sean Carroll states Ludwig Boltzmann's concept of entropy as "a measure of the number of particular microscopic arrangements of atoms that appear indistinguishable from a macroscopic perspective." Carroll, *From Eternity to Here*, Note 1 above, p. 37. Chapter 8 provides an in-depth discussion.

3. For a discussion of Hubble's discovery, see Krauss, Lawrence M., *A Universe From Nothing: Why There is Something Rather Than Nothing*, Free Press 2012, pp. 6–9.

4. See "GN-z11" at <https://en.wikipedia.org/wiki/GN-z11> (Accessed August 8, 2018). Proper distance takes into account the expansion of the universe while the light from the galaxy is approaching us.

5. See Kaku, Michio, *Einstein's Cosmos: How Albert Einstein's Vision Transformed Our Understanding of Space and Time*, Atlas Books 2004, pp. 59–67. See also Bodanis, David, *E=mc²: A Biography of the World's Most Famous Equation*, Walker Publishing Company, Inc. 2000, pp. 81–84.

6. Einstein's theory of special relativity employs Lorentz transformations, which involve basic operations of square roots and fractions. See, e.g., Halpern, Paul, *The Great Beyond: Higher Dimensions, Parallel Universes, and the Extraordinary Search for a Theory of Everything*, John Wiley & Sons, Inc. 2004, pp. 70–83, also discussing the relationship to art. For a mathematical derivation, see Collier, Peter, *A Most Incomprehensible Thing: Notes Towards a Very Gentle Introduction to the Mathematics of Relativity*, Incomprehensible Books 2014, Sec. 3.4.

7. Greene, Brian, *The Elegant Universe: Superstrings, Hidden Dimensions, and the Quest for the Ultimate Reality*, Vintage Books 2000, p. 42.

8. See JC Hafele and Richard E. Keating, "Around-the-World Atomic Clocks: Observed Relativistic Time Gains," *Science*, July 14, 1972, pp. 168–70, available at <http://science.sciencemag.org/content/177/4044/168> (Accessed August 8, 2018).

9. Greene, Brian, *The Hidden Reality: Parallel Universes and the Deep Laws of the Cosmos*, Alfred A. Knopf 2011, p. 15.

Chapter 6

1. Holt, Jim, *Why Does the World Exist: An Existential Detective Story*, Liveright Publishing Corp. 2012, p. 27.

2. Holt, *Why Does the World Exist*, Note 1 above, p. 12.

3. See, e.g., Krauss, Lawrence M., *A Universe From Nothing: Why There is Something Rather Than Nothing*, Free Press 2012.

4. See, e.g., Chris Taylor, "Elon Musk basically thinks we're living in 'The Sims.' Here's why that's wrong—

and dangerous," Mashable, June 2, 2016, at <https://mashable.com/2016/06/02/musk-computer-simulation/#I.rYcUxUfPqH> (Accessed August 8, 2018).

5. See Sean Carroll blog, "Maybe We Do Not Live in a Simulation: The Resolution Conundrum," August 22, 2016, at <http://www.preposterousuniverse.com/blog/2016/08/22/maybe-we-do-not-live-in-a-simulation-the-resolution-conundrum/> (Accessed August 8, 2018).

6. Nick Bostrom, "Are We Living in a Computer Simulation?" *Philosophical Quarterly* (2003), Volume 53, Issue 211, pp. 243–255.

7. Jason Koebler, "Elon Musk Says There's a 'One in Billions' Chance Reality Is Not a Simulation," Motherboard, June 2, 2016, at <https://motherboard.vice.com/en_us/article/8q854v/elon-musk-simulated-universe-hypothesis> (Accessed August 8, 2018).

Chapter 7

1. Dawkins, Richard, *The Selfish Gene: 40th Anniversary edition (Oxford Landmark Science)*, Oxford University Press 2016, pp. 180–81.

2. Darwin, Charles, *On the Origin of Species by Means of Natural Selection, or the Preservation of Favoured Races in the Struggle for Life*, John Murray, publisher, 1859.

3. See Michael Marshall, "Timeline: The evolution of life," *NewScientist*, July 14, 2009, at <https://www.newscientist.com/article/dn17453-timeline-the-evolution-of-life/> (Accessed August 8, 2018).

4. "Earth's Calendar Year: 4.5 billion years compressed into 12 months," Biomimicry 3.8, 2016, at <https://biomimicry.net/earths-calendar-year-4-5-billion-years-compressed-into-12-months/> (Accessed August 8, 2018).

5. Some of the extraordinary milestones in the path to human evolution are discussed in Livio, Mario, *The Accelerating*

Universe: Infinite Expansion, the Cosmological Constant, and the Beauty of the Cosmos, John Wiley & Sons, Inc. 2000, pp. 217–20.

6. See generally Omoto, Charlotte K., *Genes and DNA: A Beginner's Guide to Genetics and Its Applications,* Columbia University Press 2012, especially Chapter 4.

7. Nicholas Bakalar, "The Smiling Axolotl Hides a Secret: A Giant Genome," *New York Times,* February 1, 2018.

8. Omoto, *Genes and DNA,* Note 6 above, pp. 22–25 (e-book version).

9. Omoto, *Genes and DNA,* Note 6 above, pp. 44–45 (e-book version).

10. See, e.g., Scheck, Barry, et al., *Actual Innocence: Five Days to Execution, and Other Dispatches from the Wrongly Convicted,* Doubleday, 2000.

11. Dawkins, *The Selfish Gene,* Note 1 above. This part of Chapter 7 draws heavily from the insights of Dawkins' book.

12. Omoto, *Genes and DNA,* Note 6 above, p. 161 (e-book version).

13. For a discussion of a more machine-like future, see Paul, Gregory S. and Earl D. Cox, *Beyond Humanity: CyberEvolution and Future Minds,* Charles River Media, Inc. 1996; see also Kurzweil, Ray, *The Age of Spiritual Machines: When Computers Exceed Human Intelligence,* Viking 1999.

Chapter 8

1. Randall, Lisa, *Knocking on Heaven's Door: How Physics and Scientific Thinking Illuminate the Universe and the Modern World,* HarperCollins Publishers 2011, p. xxi.

2. "Atomic nucleus," Wikipedia <https://en.wikipedia.org/wiki/Atomic_nucleus> (Accessed August 8, 2018).

3. See Ferris, Timothy, *The Whole Shebang: A State-of-the-Universe(s) Report,* Simon & Schuster 1997, pp. 104–06.

4. Kenneth Chang, "It Was a Universe-Shaking Announcement.

But What Is a Neutron Star Anyway," *New York Times*, October 16, 2017. See also Singh, Simon, *Big Bang: The Origin of the Universe*, Harper Perennial 2004, pp. 385–90, for a discussion of how heavier elements are formed.

5. Regarding the Standard Model, see generally Randall, *Knocking on Heaven's Door*, Note 1 above. For further discussion on quarks and gluons, see Wilczek, Frank, *The Lightness of Being: Mass, Ether, and the Unification of Forces*, Basic Books 2009, Chapter 6.

6. Adrian Cho, "Once Again, Physicists Debunk Faster-Than-Light Neutrinos," *Science* magazine, June 8, 2012, at <http://www.sciencemag.org/news/2012/06/once-again-physicists-debunk-faster-light-neutrinos> (Accessed August 8, 2018). In a recent breakthrough, a neutrino was detected from the radiation surrounding a black hole in a distant galaxy. Sarah Kaplan, "In a cosmic first, scientists detect 'ghost particles' from a distant galaxy," *The Washington Post*, July 12, 2018, <https://www.washingtonpost.com/news/speaking-of-science/wp/2018/07/12/in-a-cosmic-first-scientists-detect-ghostly-neutrinos-from-a-distant-galaxy/?utm_term=.61057ed47a9f> (Accessed August 8, 2018).

7. See, e.g., Greene, Brian, *The Elegant Universe: Superstrings, Hidden Dimensions, and the Quest for the Ultimate Reality*, Vintage Books 2000, Chapter 1.

8. The story of the search for the Higgs boson is told in Carroll, Sean, *The Particle at the End of the Universe: How the Hunt for the Higgs Boson Leads Us to the Edge of a New World*, Dutton 2012.

9. Randall, *Knocking on Heaven's Door*, Note 1 above.

10. Carroll, *The Particle at the End of the Universe*, Note 8 above.

11. Carroll, *The Particle at the End of the Universe*, Note 8 above, p. 25 (e-book version).

12. Nobelprize.org <https://www.nobelprize.org/nobel_prizes/physics/laureates/2013/> (Accessed August 8, 2018).

Chapter 9

1. Feynman, Richard, *The Character of Physical Law*, BBC/Penguin Press 1965, p. 129.

2. Hobson, Art, *Tales of the Quantum: Understanding Physics' Most Fundamental Theory*, Oxford University Press 2016, p. 104 (e-book version).

3. Kaku, Michio, *Einstein's Cosmos: How Albert Einstein's Vision Transformed Our Understanding of Space and Time*, Atlas Books 2004, Chapter 3.

4. Kaku, *Einstein's Cosmos*, Note 3 above, pp. 68–69.

5. For a discussion of the important role of fields throughout science, see Brooks, Rodney A., *Fields of Color: The theory that escaped Einstein*, Epic Publications 2010.

6. Halpern, Paul, *Einstein's Dice and Schrödinger's Cat: How Two Great Minds Battled Quantum Randomness to Create a Unified Theory of Physics*, Basic Books 2015. For a mathematical derivation of Schrödinger's Equation, see Landshoff, Peter et al., *Essential Quantum Physics*, Cambridge University Press 1997, Chapter 2.

7. See Hobson, Art, *Tales of the Quantum: Understanding Physics' Most Fundamental Theory*, Oxford University Press 2016, pp. 117–20, offering a mathematical analysis of the principle. Hobson calls the limitation the Heisenberg Indeterminacy Principle.

8. See generally Musser, George, *Spooky Action at a Distance: The Phenomenon That Reimagines Space and Time—and What It Means for Black Holes, the Big Bang, and Theories of Everything*, Scientific American/Farrar, Straus and Giroux 2015.

Chapter 10

1. Greene, Brian, *The Elegant Universe: Superstrings, Hidden Dimensions, and the Quest for the Ultimate Reality*, Vintage Books 2000, p. 4.

2. "Moore's law," Wikipedia, at <https://en.wikipedia.org/ wiki/Moore%27s_law> (Accessed August 8, 2018).

3. See Barrow, John D., *The Book of Nothing: Vacuums, Voids, and the Latest Ideas about the Origins of the Universe*, Pantheon Books 2000, Chapter 1.

4. For a thorough discussion of the theory behind quantum computers, including the example used here, see generally Johnson, George, *A Shortcut Through Time: The Path to the Quantum Computer*, Alfred A. Knopf 2003, and particularly pp. 48–49. For an historical approach to the development of quantum computers, see Gribbin, John, *Computing with Quantum Cats: From Colossus to Qubits*, Prometheus Books 2014.

5. For an update on the progress in making quantum computers, see Will Knight, "Serious quantum computers are finally here. What are we going to do with them?" MIT Technology Review, February 21, 2018, at <https:// www.technologyreview.com/s/610250/serious-quantum-computers-are-finally-here-what-are-we-going-to-do-with-them/> (Accessed August 8, 2018).

6. For a discussion of quantum logic gates, see Siegfried, Tom, *The Bit and the Pendulum: From Quantum Computing to M Theory—The New Physics of Information*, John Wiley & Sons, Inc. 2000, pp. 88–91.

7. Greene, *The Elegant Universe*, Note 1 above, especially Chapter 5.

8. See Zukav, Gary, *The Dancing Wu Li Masters: An Overview of the New Physics*, HarperCollins Publishers 1979, pp. 251–61.

9. See Greene, Brian, *The Fabric of the Cosmos: Space, Time, and the Texture of Reality*, Alfred A. Knopf 2004, Chapter 12.

Bibliography

Al-Khalili, Jim. *Quantum: A Guide For the Perplexed.* Weidenfeld & Nicolson 2012

Barrow, John D. *The Book of Nothing: Vacuums, Voids, and the Latest Ideas about the Origins of the Universe.* Pantheon Books 2000

Bartusiak, Marcia. *Black Hole: How an Idea Abandoned by Newtonians, Hated by Einstein, and Gambled On by Hawking Became Loved.* Yale University Press 2015

Bean, John Gilbert. *Higgs Boson, Origin of Big Bang: The Universe, Space, And Beyond... Cosmology, Inter-Universe Travel.* John Gilbert Bean Publishing 2014

Bodanis, David. *E=mc²: A Biography of the World's Most Famous Equation.* Walker Publishing Company, Inc. 2000

Brockman, John. *The Universe: Leading Scientists Explore the Origin, Mysteries, and Future of the Cosmos (Best of Edge Series).* Harper Perennial 2014

Brooks, Rodney A. *Fields of Color: The theory that escaped Einstein.* Epic Publications 2010

Bryson, Bill. *A Short History of Nearly Everything.* Broadway Books 2003

Burdick, Alan. *Why Time Flies: A Mostly Scientific Investigation.* Simon & Schuster; Reprint edition 2017

Carroll, Sean. *The Particle at the End of the Universe: How the Hunt for the Higgs Boson Leads Us to the Edge of a New World.* Dutton 2012

Carroll, Sean. *The Big Picture: On the Origins of Life, Meaning, and the Universe Itself.* Dutton 2016

Carroll, Sean. *From Eternity to Here: The Quest for the Ultimate Theory of Time.* Dutton 2010

Collier, Peter. *A Most Incomprehensible Thing: Notes Towards a Very Gentle Introduction to the Mathematics of Relativity.*

Incomprehensible Books 2014

Dawkins, Richard. *The Selfish Gene: 40th Anniversary edition (Oxford Landmark Science)*. Oxford University Press 2016

Farmelo, Graham. *The Strangest Man: The Hidden Life of Paul Dirac, Mystic of the Atom*. Basic Books 2009

Ferris, Timothy. *Seeing in the Dark: How Backyard Stargazers Are Probing Deep Space and Guarding Earth from Interplanetary Peril*. Simon & Schuster 2002

Ferris, Timothy. *The Mind's Sky: Human Intelligence in a Cosmic Context*. Bantam Books 1992

Ferris, Timothy. *The Red Limit: The Search for the Edge of the Universe*. Perennial 1997

Ferris, Timothy. *Coming of Age in the Milky Way*. Anchor Books 1998

Ferris, Timothy. *The Whole Shebang: A State-of-the-Universe(s) Report*. Simon & Schuster 1997

Ferris, Timothy, ed. *The World Treasury of Physics, Astronomy, and Mathematics*. Little, Brown and Company 1991

Frank, Adam. *About Time: Cosmology and Culture at the Twilight of the Big Bang*. Free Press; Reprint edition 2011

Gell-Mann, Murray. *The Quark and the Jaguar: Adventures in the Simple and the Complex*. WH Freeman and Company 1994

Greene, Brian. *The Elegant Universe: Superstrings, Hidden Dimensions, and the Quest for the Ultimate Theory*. Vintage Books 2000

Greene, Brian. *The Fabric of the Cosmos: Space, Time, and the Texture of Reality*. Alfred A. Knopf 2004

Greene, Brian. *The Hidden Reality: Parallel Universes and the Deep Laws of the Cosmos*. Alfred A. Knopf 2011

Gribbin, John. *In Search of the Big Bang*. ReAnimus Press; 2nd edition 2015

Gribbin, John. *Before the Big Bang*. Kindle Edition 2015

Gribbin, John. *Computing with Quantum Cats: From Colossus to Qubits*. Prometheus Books 2014

Halpern, Paul. *Einstein's Dice and Schrödinger's Cat: How Two Great Minds Battled Quantum Randomness to Create a Unified Theory of Physics.* Basic Books 2015

Halpern, Paul. *The Great Beyond: Higher Dimensions, Parallel Universes, and the Extraordinary Search for a Theory of Everything.* John Wiley & Sons, Inc. 2004

Hawking, Stephen and Leonard Mlodinow. *The Grand Design.* Bantam Books 2010

Hawking, Stephen et al. *The Future of Spacetime.* WW Norton & Co. 2002

Hazen, Robert M. *Why Aren't Black Holes Black? The Unanswered Questions at the Frontiers of Science.* Anchor Books 1997

Heeren, Fred. *Show Me God: What the Message from Space Is Telling Us About God.* Day Star Publications 1998

Hobson, Art. *Tales of the Quantum: Understanding Physics' Most Fundamental Theory.* Oxford University Press 2016

Holt, Jim. *Why Does the World Exist: An Existential Detective Story.* Liveright Publishing Corp. 2012

Johnson, George. *A Shortcut Through Time: The Path to the Quantum Computer.* Alfred A. Knopf 2003

Kaku, Michio. *Parallel Worlds: A Journey Through Creation, Higher Dimensions, and the Future of the Cosmos.* Anchor; Reprint edition 2006

Kaku, Michio. *Einstein's Cosmos: How Albert Einstein's Vision Transformed Our Understanding of Space and Time.* Atlas Books 2004

Krauss, Lawrence M. *A Universe From Nothing: Why There is Something Rather Than Nothing.* Free Press 2012

Krauss, Lawrence M. *Quantum Man: Richard Feynman's Life in Science.* Atlas Books 2011

Kurzweil, Ray. *The Age of Spiritual Machines: When Computers Exceed Human Intelligence.* Viking 1999

Landshoff, Peter et al. *Essential Quantum Physics.* Cambridge University Press 1997

Lightman, Alan. *The Accidental Universe: The World You Thought You Knew.* Vintage 2014

Lindley, David. *Where Does the Weirdness Go? Why Quantum Mechanics Is Strange But Not As Strange As You Think.* Basic Books 1996

Livio, Mario. *The Accelerating Universe: Infinite Expansion, the Cosmological Constant, and the Beauty of the Cosmos.* John Wiley & Sons, Inc. 2000

Muller, Richard A. *Now: The Physics of Time.* WW Norton & Company 2016

Musser, George. *Spooky Action at a Distance: The Phenomenon That Reimagines Space and Time—and What It Means for Black Holes, the Big Bang, and Theories of Everything.* Scientific American/ Farrar, Straus and Giroux 2015

New Scientist. *The Unknown Universe: New Scientist: The Collection.* Reed Business Information Ltd; 2nd edition 2014

Newton, Roger G. *Thinking about Physics.* Princeton University Press 2000

Omoto, Charlotte K. *Genes and DNA: A Beginner's Guide to Genetics and Its Applications.* Columbia University Press 2012

Pagels, Heinz R. *The Cosmic Code: Quantum Physics as the Language of Nature (Dover Books on Physics).* Dover Publications; Reprint edition 2012

Paul, Gregory S. and Earl D. Cox. *Beyond Humanity: CyberEvolution and Future Minds.* Charles River Media, Inc. 1996

Penrose, Roger. *Cycles of Time: An Extraordinary New View of the Universe.* Vintage 2011

Randall, Lisa. *Knocking on Heaven's Door: How Physics and Scientific Thinking Illuminate the Universe and the Modern World.* HarperCollins Publishers 2011

Reinking, Greg F. *Cosmic Legacy: Space, Time, and the Human Mind.* Vantage Press 2003

Rosenblum, Bruce et al. *Quantum Physics of Consciousness.* Cosmology Science Publishers 2011

Rovelli, Carlo. *Seven Brief Lessons on Physics*. Riverhead Books 2016

Scharf, Caleb. *The Copernicus Complex: Our Cosmic Significance in a Universe of Planets and Probabilities*. Scientific American/ Farrar, Straus and Giroux 2014

Schumacher, Benjamin. *Quantum Mechanics: The Physics of the Microscopic World*. The Great Courses 2009

Siegfried, Tom. *Strange Matters: Undiscovered Ideas at the Frontiers of Space and Time*. Joseph Henry Press 2002

Siegfried, Tom. *The Bit and the Pendulum: From Quantum Computing to M Theory—The New Physics of Information*. John Wiley & Sons, Inc. 2000

Singh, Simon. *Big Bang: The Origin of the Universe*. Harper Perennial 2004

Stone, A. Douglas. *Einstein and the Quantum: The Quest of the Valiant Swabian*. Princeton University Press; Reprint edition 2015

Susskind, Leonard. *The Black Hole War: My Battle With Stephen Hawking to Make The World Safe for Quantum Mechanics*. Little, Brown and Company 2008

Susskind, Leonard and Art Friedman. *Quantum Mechanics: The Theoretical Minimum*. Basic Books 2014

Thorne, Kip. *The Science of Interstellar*. WW Norton & Company 2014

Unger, Roberto Mangabeira and Lee Smolin. *The Singular Universe and the Reality of Time: A Proposal in Natural Philosophy*. Cambridge University Press 2014

Weatherall, James Owen. *Void: The Strange Physics of Nothing (Foundational Questions in Science)*. Yale University Press 2016

Wilczek, Frank. *The Lightness of Being: Mass, Ether, and the Unification of Forces*. Basic Books 2009

Wilson, Edward O. *The Meaning of Human Existence*. Liveright 2014

Yourgrau, Palle. *Gödel Meets Einstein: Time Travel in the Gödel*

Universe. Open Court Publishing Company 1999

Zukav, Gary. *The Dancing Wu Li Masters: An Overview of the New Physics*. HarperCollins Publishers 1979

ACADEMIC AND SPECIALIST

Iff Books publishes non-fiction. It aims to work with authors and titles that augment our understanding of the human condition, society and civilisation, and the world or universe in which we live.

If you have enjoyed this book, why not tell other readers by posting a review on your preferred book site.

Is There an Afterlife?

David Fontana

Is there an Afterlife? If so what is it like? How do Western ideas of the afterlife compare with Eastern? David Fontana presents the historical and contemporary evidence for survival of physical death.

Paperback: 978-1-90381-690-5

Nothing Matters

a book about nothing

Ronald Green

Thinking about Nothing opens the world to everything by illuminating new angles to old problems and stimulating new ways of thinking.

Paperback: 978-1-84694-707-0 ebook: 978-1-78099-016-3

Panpsychism

The Philosophy of the Sensuous Cosmos

Peter Ells

Are free will and mind chimeras? This book, anti-materialistic but respecting science, answers: No! Mind is foundational to all existence.

Paperback: 978-1-84694-505-2 ebook: 978-1-78099-018-7

Punk Science

Inside the Mind of God

Manjir Samanta-Laughton

Many have experienced unexplainable phenomena; God, psychic abilities, extraordinary healing and angelic encounters. Can cutting-edge science actually explain phenomena previously thought of as 'paranormal'?

Paperback: 978-1-90504-793-2

The Vagabond Spirit of Poetry

Edward Clarke

Spend time with the wisest poets of the modern age and of the past, and let Edward Clarke remind you of the importance of poetry in our industrialized world.

Paperback: 978-1-78279-370-0 ebook: 978-1-78279-369-4

Readers of ebooks can buy or view any of these bestsellers by clicking on the live link in the title. Most titles are published in paperback and as an ebook. Paperbacks are available in traditional bookshops. Both print and ebook formats are available online.
Find more titles and sign up to our readers' newsletter at
http://www.johnhuntpublishing.com/non-fiction
Follow us on Facebook at
https://www.facebook.com/JHPNonFiction
and Twitter at https://twitter.com/JHPNonFiction